THE REVOLUTION IN MILITARY AFFAIRS

FOREIGN POLICY, SECURITY AND STRATEGIC STUDIES

Editors: Alex Macleod and Charles-Philippe David

The Foreign Policy, Security and Strategic Studies Series seeks to promote analysis of the transformation and adaptation of foreign and security policies in the post–Cold War era. The series welcomes manuscripts offering innovative interpretations or new theoretical approaches to these questions, whether dealing with specific strategic or policy issues or with the evolving concept of security itself.

MONOGRAPHS
Canada, Latin America, and the New Internationalism
A Foreign Policy Analysis, 1968–1990
Brian J.R. Stevenson

Power vs. Prudence
Why Nations Forgo Nuclear Weapons
T.V. Paul

From Peacekeeping to Peacemaking
Canada's Response to the Yugoslav Crisis
Nicholas Gammer

Canadian Policy toward Khrushchev's Soviet Union
Jamie Glazov

The Revolution in Military Affairs
Implications for Canada and NATO
Elinor Sloan

COLLECTIONS
Nato after Fifty
Enlargement, Russian, and European Security
Edited by Charles-Philippe David and Jacques Lévesque

The Revolution
in Military Affairs
Implications for Canada
and NATO

Elinor C. Sloan

The Centre for Security and Foreign Policy Studies
and
The Raoul-Dandurand Chair of Strategic and Diplomatic Studies

McGill-Queen's University Press
Montreal & Kingston · London · Ithaca

Legal deposit second quarter 2002
Bibliothèque nationale du Québec

Printed in Canada on acid-free paper.

This book has been published with the help of a grant from
the Humanities and Social Sciences Federation of Canada,
using funds provided by the Social Sciences and Humanities
Research Council of Canada.

McGill-Queen's University Press acknowledges the financial
support of the Government of Canada through the Book
Publishing Industry Development Program (BPIDP) for its
publishing activities. We also acknowledge the support of the
Canada Council for the Arts for our publishing program.

The author acknowledges the permission granted by the Director
of Intellectual Property, Department of National Defence, to use
excerpts from reports originally prepared for the Directorate of
Strategic Analysis in the preparation of this work. Although
parts of this work were originally undertaken for the Canadian
Department of National Defence, this book does not necessarily
reflect the views of the department.

**National Library of Canada Cataloguing in
Publication Data**

Sloan, Elinor C. (Elinor Camille), 1965-
    The revolution in military affairs: implications for
Canada and NATO
    (Foreign policy, security and strategic studies)
    Published for the Centre for Security and Foreign
Policy Studies and the Raoul-Dandurand Chair of
Strategic and Diplomatic Studies.
    Includes bibliographical references and index.
    ISBN 0-7735-2363-4 (bound). – ISBN 0-7735-2394-4
(pbk.)
    1. Military art and science – Technological
innovations – History – 20th century.   2. Security,
International.   3. North Atlantic Treaty
Organization – Military policy.   4. Canada –
Military policy.   I. Université du Québec à Montréal.
Centre d'études des politiques étrangères et de sécurité
II. Raoul-Dandurand Chair of Strategic and
Diplomatic Studies III. Title.   IV. Series.
U42.S56 2002   355'.0330049   C2001-903613-2

This book was typeset by Dynagram Inc. in 10.5/13 Sabon.

For Tom, Thomas, and William

Contents

Introduction

Advances in information technology are driving a high-tech revolution in military affairs (RMA) that is transforming the nature of modern warfare. New technologies are being harnessed and incorporated into new military doctrines and organizational concepts to such an extent that the ultimate result could render obsolete or irrelevant the doctrines and concepts that guided the great powers throughout the Cold War period. This process began in a limited fashion with the introduction of precision munitions in the latter stages of the Vietnam War and took root in the late 1970s and 1980s with America's "offset strategy" of countering Soviet numerical supremacy in troops with Western technological strength. The Gulf War provided the first concrete, vivid indication that a military technological revolution was under way. This conflict can be seen as the precursor to a broader RMA that, balancing the onward march of technology with the constraints of defence budgets and bureaucratic inertia, could come to fruition anytime between 2015 and 2025.

Although the RMA has been the subject of much discussion in the United States since the early 1990s and has figured in official U.S. defence policy for several years, it has not received the same level of analytical attention in Canada or elsewhere. Outside the rather limited defence policy circles there has been little discussion of the changing nature of warfare, the developments it encompasses, and the potential foreign and defence policy implications for Canada and NATO. This book seeks to bridge that gap. Its goal is to explore the diverse dimensions of the RMA, how Canada and its allies are responding to its challenges, and to what extent the RMA is applicable to the sorts of military missions to which they may want to commit their forces in the future. Taking this analysis into account, the book sets out a way ahead for Canada and NATO.

Chapter 1 begins by making the somewhat abstract phrase "revolution in military affairs" more comprehensible by outlining the key technological, doctrinal, and organizational developments that make up the RMA. Chapter 2 places the RMA in historical context, providing a brief overview of how selected thinkers have evaluated military revolutions over the course of history and, in doing so, have offered valuable insights into the current transformation in military affairs. The chapter discusses the origins of the current RMA, its strategic drivers, and some cautionary points about the promises its offers. Chapter 3 addresses the question, Is there an RMA? by examining concrete efforts that the United States, which is at the forefront of RMA-related developments, is taking to transform its military forces and thereby exploit the revolutionary potential of military technologies. Chapter 4 broadens this perspective by looking at allied approaches to the RMA, focusing on Britain, Australia, France, and Germany. Chapter 5 analyses more specifically the technology and capability gap between the United States and its NATO allies and the role the alliance's Defence Capabilities Initiative is playing in addressing this gap.

Chapter 6 examines the relevance of the RMA to peace-support operations, while chapter 7 discusses "asymmetric" threats – which were recognized even before the terrorist attacks on the United States as being the primary means by which U.S. adversaries would be likely to respond to America's pursuit of the RMA. Chapter 8 outlines measures Canada is taking to respond to the technological, doctrinal, and organizational imperatives of the RMA and highlights some important contextual factors that will frame its future approach. The concluding chapter draws on the key themes and conclusions of the preceding chapters to set out a way ahead for Canada and NATO in responding to the U.S.-led revolution in military affairs.

Although this book was on its way to press in the summer of 2001, it has been modified in preliminary fashion throughout, and particularly in chapters 3 and 7, to reflect the tragic events of

11 September 2001. The epilogue captures more extensively the relevance of revolutionary technologies and concepts to prosecuting the war against international terrorism and, conversely, the impact of the terrorist attacks on the transformation of U.S. military forces and the eventual realization of an RMA. Beyond this, it places the RMA in the context of a new era, a "post–September 11 era" in which the United States and its NATO allies are engaged once again in a political, military, and ideological struggle against a common foe and in which they must once again respond with the necessary military resources and capabilities.

THE REVOLUTION IN MILITARY AFFAIRS

CHAPTER ONE

What Is the Revolution
in Military Affairs?

Definitions of a revolution in military affairs are wide and varied and perhaps as numerous as its analysts. They range from those that capture with sweeping simplicity the essential nature of an RMA ("an RMA is simply a revolutionary change in how wars are fought and won – a change that can often be recognised by the ease with which 'participating' armed forces can defeat 'non-participating' ones")[1] to those that highlight its defining characteristics ("an RMA involves a paradigm shift in the nature and conduct of military operations which *renders obsolete* or *irrelevant* one or more *core competencies* of a dominant player")[2] to those that enunciate its specific elements (an RMA is "a major change in the nature of warfare brought about by the innovative application of technologies which, combined with dramatic changes in military doctrine and operational and organizational concepts, fundamentally alters the character and conduct of military operations").[3] The implicit or explicit statement that an RMA includes more than technological advances is common to all these definitions. Indeed, the central tenet of an RMA is that advances in technology must lead to significant changes in how military forces are organised, trained, and equipped for war, thereby reshaping the way in which wars are fought.

There have been many revolutions in military affairs over the course of history, some of which will be discussed in chapter 2. What follows here is an examination of the key technological, doctrinal, and organizational attributes of today's ongoing revolution in military affairs. Because the United States is at the forefront of most changes associated with this RMA, examples focus to a large degree on developments in the U.S. military services. Subsequent chapters will explore how other Western nations are responding to the revolution in military affairs.

REVOLUTIONARY TECHNOLOGIES

Precision Force and Precision-Guided Munitions

Perhaps the best-known technological advance in modern warfare is the development of precision-guided munitions. Such munitions were first developed in the latter stages of the Vietnam War, but their accuracy has increased dramatically since about the mid-1980s, and even more so since the Gulf War. Today's "smart" or "brilliant" weapons include a host of guided, precision, self-activated weapons that range from missiles to individual warheads to defences against enemy smart weapons. The u.s. and British navies' *Tomahawk* cruise missile, guided by the Global Positioning System (GPS), can reliably hit a target the size of a small room from a thousand miles away.[4] The u.s. Army's second generation Tactical Missile System will be able to destroy battalion-sized formations of moving armoured combat vehicles out to ranges in excess of 140 kilometres when it is loaded with the Brilliant Anti-Armour Submunition.[5] Air Force bomber pilots can drop precision bombs such as the Joint Direct Attack Munition from six miles up and 40 miles away and hit their targets despite darkness, smoke, or bad weather.[6] The u.s. Air Force plans to phase out almost all unguided bombs by 2005.[7]

Precision-guided munitions have enabled the application of "precision force," which refers more generally to the use of deadly violence with greater speed, range, and accuracy. Smart weapons have already brought an increase in destructive power by as much as a factor of one thousand over the unguided bombs of yesterday.[8] At a time when there is growing public intolerance for casualties, precision-guided munitions also offer the possibility of destroying military targets without substantial "collateral," that is, civilian, damage.

Force Projection and Stealth

Technological advances are also being made in major military platforms, increasing force-projection capabilities. Here perhaps the most significant development is in the area of "low-observable" technologies, or stealth. Stealthy platforms have an advantage over their nonstealthy counterparts because of their ability to penetrate high-threat areas and deliver precision-guided munitions. The U.S. Air Force already has a fleet of stealthy F-117 fighter aircraft, as well as a number of stealthy B-2 bombers. Looking to the future, the stealthy F-22 air-superiority fighter, the first of the next-generation U.S. combat aircraft, is in advanced development, with a planned operational date before 2005. This will be followed by the stealthy, multirole Joint Strike Fighter, which is now being developed in three variants for the U.S. Air Force, Navy, and Marine Corps, as well as for Britain's Royal Navy and potentially for the air forces of several other countries, including Canada. It is scheduled for initial deployment in 2010 to 2020. Also around 2010, the U.S. Air Force expects to field a stealthy Unmanned Combat Aerial Vehicle (UCAV).

America's nuclear-propelled submarines are already designed to be the stealthiest in the world. In addition, the navy's next-generation destroyer will be a much larger surface combatant than current designs but will have the substantially lower signature levels of submarines.[9] Leading the way in all-out surface-ship stealth design are European navies, in particular the Swedish and French navies.[10]

Stealth technologies are also being embraced by ground forces and are slowly making their way into army reequipment programs. For example, the U.S. Army's *Comanche* armed reconnaissance helicopter incorporates "stealthy" features designed to protect the aircraft from visual, radar, and thermal infrared detection. Similarly, Britain's new armoured fighting vehicle, the *Warrior 2000*, has lower thermal and acoustic signatures than its predecessor.[11]

Although stealth technology can be very useful – the F-117 proved highly effective during the Gulf War – it has its limitations. Stealth technologies require continual advances to make up for improved detection capabilities. U.S. military sources acknowledge that current stealth technology will likely be defeated in the next few years by advances in radar and infrared technologies.[12] Moreover, simply operating stealth aircraft makes them less able to cloak themselves from radar because of wear and tear to their radar-absorbing materials. Testing has indicated, for example, that the exposure of B-2 bombers to water and humidity can damage some of the low-observable-enhancing surfaces on the aircraft. Stealth aircraft also often have less endurance and payload capacity than their nonstealthy counterparts. Therefore, the advantages of stealth must be weighed against the trade-off in performance capability. Finally, the Kosovo experience of an F-117 fighter being shot down by Serbian antiaircraft fire gave concrete evidence that stealth is not invincible.

Nor are stealth ships without their problems. During trials of America's first stealth ship, the *Sea Shadow*, which was retired in 1995, designers discovered that stealth can work *too* well. A large blank area on a radar-reflecting sea gave away the position of the ship. Accordingly, stealth designers are now focusing on making smaller vessels that operate in shallow littoral waters but are still manageable on the high seas.[13]

Battlespace Awareness and Control

More than precision munitions or the contribution of any particular military platform, it is the potential of new military technologies to reduce the "fog of war" that could change the way wars are fought. Sensors in satellites, manned aircraft, or Unmanned Aerial Vehicles (UAVs) can now monitor virtually everything that is going on in a particular battle area, dramatically improving battlespace awareness. The net result is the potential for commanders to have complete, real-time knowledge of the disposition of all enemy and friendly forces.

U.S. Defense Support Program satellites, originally designed to detect intercontinental ballistic missiles, have been improved to the extent that they are now able to spot and analyse theatre missile launches and transmit a warning in time to help troops in danger. Improvements in the GPS satellite navigation network, which now has twenty-eight satellites, compared to sixteen during the Gulf War, have dramatically improved space-based sensor capabilities. Depending on the nature of the terrain, America's Joint Surveillance Target Attack Radar System (JSTARS), an airborne ground surveillance system, can display the position of vehicles in any weather within an area of two hundred square kilometres. JSTARS upgrades will soon enable these aircraft to locate, track, and identify vehicles even in mountainous terrain. American and NATO Airborne Warning and Control System (AWACS) aircraft provide surveillance capabilities similar to the capabilities of JSTARS aircraft, but for airspace.

In the arsenals of many Western militaries tactical UAVs are already a staple item for reconnaissance, surveillance, and electronic warfare. The *Pioneer* is a shipboard UAV developed by Israel in the 1980s to provide imagery intelligence for tactical commanders on land and at sea. This system, as well as America's smaller *Exdrone* and *Pointer* systems, was used successfully to gather intelligence data during the Gulf War of 1990–91. The *Pioneer* also supported operations in Bosnia, Haiti, and Somalia, while the U.S. Army's *Hunter* tactical UAV made its combat debut during NATO's operation in and around Kosovo in 1999. Late in 2000 the army began fielding its *Shadow* tactical UAV.

Since the Gulf War there has been increasing interest in developing UAVs for strategic roles. Capable of flying autonomously at high altitudes for long periods of time, strategic UAVs provide continuous reconnaissance over broad stretches of territory. America's medium-altitude, long-endurance *Predator* UAV can loiter on station for twenty-four hours at four hundred nautical miles from its launch point and at altitudes up to twenty-five thousand feet. It can provide specific imagery within fifteen minutes of a request, compared to about thirty-six hours for a manned reconnaissance aircraft. *Predator* UAVs were used extensively during the air campaign in Kosovo in 1999, and during the war on terrorism in Afghanistan in 2001. Since 1996 they have been monitoring compliance with the Bosnian Peace Accord. America's one-ton payload, high-altitude *Global Hawk* UAV was also used in Afghanistan. It can fly three thousand nautical miles at up to sixty-seven thousand feet and loiter over a target area for twenty-four hours at a time.

The U.S. Army is enhancing its battlespace awareness by "digitizing" the battlefield. That is to say, digital technology is being added to or built into aircraft, tanks, artillery, and individual soldier systems, with the intention of providing commanders with an instantaneous and complete picture of the battlefield. Each soldier and vehicle is to be equipped with a small computer that displays a map of the battlefield overlaid with friendly and enemy positions and aircraft flight paths. The army fielded it first digitized division in 2000, the Fourth Infantry Division (Mechanized), and plans to field its first digitized corps by October 2004.

Enhancing battlespace awareness is only part of the story; equally important as cutting through the "fog of war" is having the command and control architecture to act on information. Advanced command, control, communications, computers and intelligence processing (C4I) systems are being designed to make sense of the vast amount of data that is gathered, display it in a useful fashion on screen, and assign targets to missiles and tanks. They are also meant to enable commanders to be in constant and instant contact with every subordinate element of the force. By reducing the surveillance-synthesis-assessment-command-strike loop to a matter of minutes, advanced command and control systems have the potential to allow a commander to control a battle from one moment to the next. They could eventually act as a "force multiplier," in that dominance in this area alone could yield war-winning advantages.

Not surprisingly, all three U.S. services are focusing on advanced command and control systems. The U.S. Navy is exploiting such systems in the context of its "network-centric warfare" initiative (discussed below). The U.S. Air Force has developed a Joint Situational Awareness System, which, unlike previous command and control systems, merges battlefield intelligence gathered by spy satellites, commercial satellites, U-2 reconnaissance aircraft, AWACS, JSTARS, and *Predator* unmanned surveillance aircraft.[14] And, as part of its digitization effort, the U.S. Army is putting in place a Battle Command Brigade and Below tactical command and control system designed to pass digital data through the army's tactical internet, thereby providing commanders with real-time situational awareness and command and control.[15]

The ultimate objective of advanced battlespace awareness and control capabilities is "information dominance," defined by the U.S. Army as the "delta" between what friendly forces know and what the enemy knows.[16] During the Gulf War, the U.S.-led coalition enjoyed an information advantage as a result of a communications network that linked satellites, observation aircraft, and commanders. Nonetheless, the numerous, sophisticated reconnaissance, communications, and targeting systems still had difficulty communicating with one another. To address this problem, in 1995 U.S. Admiral William Owens outlined a vision of a "system of systems" in which systems that provide for battlespace awareness, the use of precision force, and advanced C4I would use more compatible software and hardware to increase their ability to work together. These technological advances are in the process of being implemented in the United States.

REVOLUTIONARY DOCTRINES

Joint Doctrine

Most military experts believe the new military will increasingly be a joint force. "Technology is forcing a long-overdue movement toward true combined-arms operations," argues one RAND analyst.[17] Indeed, the RMA is bringing about an increasingly integrated battlefield, with the army, navy and air force working ever more closely together. Future scenarios might see air force precision force preparing the battlefield for ground forces and airlift assets transporting troops to the theatre of operations. Manned, unmanned, and satellite surveillance platforms would operate throughout the campaign, supporting all three services, while naval forces could provide offshore logistical support, sea lift, and precision force capabilities against ground targets.

The RMA also foresees greater combined operations potential among the armed forces of different countries. While "jointness" refers to increased operational integration among the various components of the armed services, "combined" operations involve the military services working in coalition with their counterparts from other countries. This trend towards joint and combined operations will place a premium on measures among services and militaries to ensure interoperability – defined as the ability of systems, units, and forces to provide services to and accept services from other systems, units, or forces and to use these services to enable them to operate effectively together.

Naval Doctrine

It follows that a key shift in naval doctrine is towards littoral warfare. In contrast to the Cold War scenario of battling against a large, opposing surface force at sea, naval experts expect that most future contingencies will require the navy to project force "from the sea," directly ashore, whether in the context of a regional war or a peacekeeping operation.[18] Accordingly, the U.S. Navy's most recent strategy statement, *Forward ... From the Sea*, stresses that the navy must move from preparing to fight on the oceans to projecting power from the sea to the land, in support of marine or army units. To this end, the navy has transformed its *Tomahawk* land-attack cruise missile from a strategic weapon into one that can also strike tactical targets ashore. It is also developing the DD-21 Land Attack Destroyer to provide precision firepower in support of combat soldiers and marines, with an expected fleet introduction date of about 2010. The littoral region is a perfect battleground for joint warfare, because effective operations in this environment require the navy to work closely with the army.

A second important doctrinal change within the U.S. Navy is a shift from platform-centric to network-centric warfare. This concept places the emphasis on the sensor and surveillance systems of a group of warships, submarines, or aircraft, rather than on the particular attributes of the platform itself.[19] Critical to this approach is the timely exchange of data among many diverse platforms, in order to reduce reaction time to an absolute minimum and deny the enemy a window to respond. The Navy's Co-operative Engagement Capability system is designed to pass sensor data so quickly within a battle group that one warship will be able to shoot down an incoming missile by using another ship's radar picture. During the coming decade, the U.S. Navy's *Aegis* air defence ships and E-2C *Hawkeye* airborne early-warning aircraft will become fully networked via the system, enabling them to transmit and receive target data to and from each other and fuse this information into a common, much more accurate image.

Land Doctrine

Land warfare specialists argue that regardless of the doctrinal changes taking place in the navy and air force (see below), it still takes "boots on the ground" to achieve U.S. military objectives. Most of the military missions that the United States has faced since the end of the Cold War have depended on ground forces backed by airlift, sealift, tactical air power, logistics, communications, and intelligence. In Bosnia, for example, aircraft carriers, marines, and fighter aircraft could not compensate for the multinational army divisions that were needed to implement the peace. Similarly, following its air campaign in and around Kosovo NATO deployed a ground force to secure the peace and facilitate the return and resettlement of over eight hundred thousand refugees. Even during the Gulf War, Iraqis endured being fired upon from remote locations but did not surrender until American and coalition ground forces were present on the battlefield. And in the war on terrorism in Afghanistan, special operations forces played an integral role.

Although ground forces will continue to be necessary, their characteristics must change. The unpredictable nature of the threats in today's international security environment necessitates that military forces have the ability to respond quickly to almost any situation. This, in turn, demands smaller, more mobile and flexible ground forces that are still highly lethal. The idea is to change from a forward-deployed industrial age army trained, equipped, and postured to stop a Soviet advance in Europe to an information age power-projection army. To meet this need, several countries have developed or plan to develop rapid reaction or expeditionary forces. The United States Army, for example, is in the process of transforming itself into a rapidly deployable, medium-weight strike force (see chapter 3), and at its 1999 summit in Helsinki the European Union declared its intention to field a rapid reaction force by 2003 (see chapter 5).

An expeditionary deployment capability depends in part on reorganising armies into smaller, more modular and mission-tailorable units. Ground forces also require lighter and more mobile (especially air deployable) equipment that remains highly lethal and does not sacrifice force protection. The overall trend in many Western countries today is to develop armoured fighting vehicles that move away from the average seventy-ton weight during the Cold War and toward a thirty-to forty-ton range – or even less. The U.S. Army has placed a very ambitious twenty-ton weight limit on its Future Combat System family of vehicles. In addition, the U.S. Congress has mandated that a third of the U.S. services' ground combat vehicles be remotely operated by 2015. Finally, an effective rapid reaction capability is dependant on the availability of strategic lift, and especially airlift, since this is the only mode of transport that can react to global force-commitment needs within days. Not surprisingly the market for long-range military transport aircraft, such as America's

C-17 heavy-lift aircraft and Europe's A400M Future Large Aircraft, has grown especially active.

The smaller, more mobile forces of tomorrow must still be able to "pack a punch." By tying enhanced reconnaissance and surveillance systems to precision-guided munitions, the combat capabilities of army units will be multiplied. Information age technologies will confer greater warfighting capabilities on smaller combat formations, giving them the strength of units many times larger and eliminating the need for large troop build-ups in the conflict area.[20] The idea of massing troops in conflict will thus be traded for that of massing firepower and carefully selecting the time and place of attack.

Future ground operations are likely to see troops moving rapidly over the battlefield in small groups.[21] These forces will be "nonlinear" in the sense that they will be more widely dispersed than in the past and will bear little resemblance to the front line of yesterday. Increased dispersion and mobility will be possible offensively, because each platform or individual soldier will carry weapons with greater lethality and reach. Synchronization of the dispersed elements will be critical, placing a premium on communications and coordination capabilities. Helicopters will be important for enhancing battlefield mobility and lethality. "Crisis management will proceed less from the military capacity to crush any opponent and more from the ability to quickly reduce the ambiguity of a situation, to respond flexibly, and to use force, where necessary, with precision and accuracy."[22]

To transform themselves into the smaller, more mobile force of the future, land forces must become less dependent on large logistics infrastructures. A primary reason why weapons systems have remained embedded in large, unwieldy units is that it has only been in such organizations that the systems could obtain the proper supplies, spare parts, and repairs.[23] To overcome this factor, the United States has developed the concept of "focused logistics," which results from applying advanced technologies to logistics efforts. The Pentagon uses navigational systems such as the Global Positioning System, the Inertial Navigation System, and the Enhanced Position Location System to track supplies around the world, knowing the contents, location, and time of arrival of each container.[24] This is allowing U.S. forces to shift from a so-called just-in-case logistics system to just-in-time delivery, eliminating the need for large stockpiles.

In their efforts to become more mobile and flexible, land forces are also purchasing equipment that provides more combat capability while requiring less logistics support. For example, precision-guided munitions indirectly affect logistics, because better weapons accuracy reduces the need for extensive ammunition supplies. Experts predict that moving to next-generation munitions will allow armies to buy fewer systems and reduce stockpiles even further.[25]

Air Doctrine

Air-power proponents argue that the ability of modern air power to affect land warfare has crossed a threshold over which its effects are fundamentally greater than before. Advances in precision capabilities have made air power the decisive force in war, allowing for the doctrine of "disengaged combat," under which ground commanders do not have to face the enemy directly until the costs of such contact have been made tolerable.[26] In high-intensity combat, air power proponents argue, the principal role of land power may now be to secure a win, rather than achieve it.[27] A related concept is that of "rapid halt," according to which an immediate and overwhelming application of air power against an attacking enemy can stop an assault, disrupt the enemy's ability to control its forces, paralyse the assailant, and pave the way for an easier ground campaign.[28] As with ground forces, air forces are adopting expeditionary doctrines. To transform itself from a Cold War force structured to contain the Soviet Union to one focused on countering the full spectrum of conflict, the U.S. Air Force has reestablished itself as an "expeditionary aerospace force" (see chapter 3).

Many military experts predict a long-term move away from manned fighters and toward unmanned combat. "The transition from operating UAVs as sensor platforms to employing them as weapons carriers is a major step, but it will surely come," argues one analyst.[29] As a means of bridging the gap before its stealthy UCAV can be fielded, the U.S. Air Force has armed some of its *Predator* UAVs with precision weapons. They were used during the war on terrorism in Afghanistan. The U.S. Congress has mandated that a third of all U.S. deep-strike aircraft be unmanned by the end of this decade.[30]

Under a doctrine of unmanned combat, pilots sitting in an execution centre in the United States would manoeuvre UCAVs through satellite communication links. Such vehicles would have a number of advantages over their manned counterparts. They would be significantly less expensive, they would remove the risk of aircrew casualties and, in some cases, they might even be able to outperform manned aircraft. Whereas a pilot in a modern fighter is limited to a maximum gravitational force of 9g, due to the effects of g-induced loss of consciousness, unmanned aircraft can withstand up to 20g of force. As a result, unmanned tactical aircraft would be able to travel substantially faster, higher, and longer than would manned fighters. A 1993 study by Lockheed Martin demonstrated that through aggressive manoeuvring, unmanned aircraft could out-turn defensive fighters and even their missile armament.[31] Taking the man out of the machine would also allow for increased stealth, since without a pilot the aircraft could be made substantially smaller. A 1996 study by the Scientific Advisory Board concluded that UCAVs have the potential to accomplish tasks that are difficult or risky for manned aircraft, such as the suppression of enemy air defences.[32]

These factors raise the question of whether and for how long manned aircraft will be needed for the delivery of precision-guided munitions. It is conceivable that in the future manned fighters may be used for only a limited number of roles, such as shooting down the most advanced enemy aircraft. At a minimum, stealthy unmanned combat aircraft, together with low-observable long-range stand-off munitions, will likely lessen the need for manned aircraft to penetrate enemy defences. Boeing has already proposed developing a UCAV with the Joint Strike Fighter's capabilities by 2020.

Space

Several military experts argue that space will evolve in the future from a medium for military and commercial satellite transmission into an actual theatre of military operations. "If historical patterns are any guide," one analyst has argued, "the coming military revolution will witness the militarization of space, with warfare occurring in space as well as on land, at sea and in the air."[33] Space operations could involve everything from protecting military satellites to knocking out enemy space-borne threats and denying adversaries the same opportunities in space. Future threats to satellite systems could include satellites armed with lasers, as well as electronic jamming devices and viruses that could shut down the flow of information. To counter such threats, the u.s. Army has developed and tested a ground-based antisatellite weapon involving a highly powerful laser. At the same time, the air force is carrying out a navigation-warfare advanced-concept-technology demonstration designed to develop measures for ensuring unimpeded access to the GPS by American and allied forces.

REVOLUTIONARY ORGANIZATIONS

It is not only the tools and doctrines of warfare but the organizations that wield them that make for revolutionary change in war. Military experts often note that the invention of the tank did not bring about an armoured warfare revolution until Germany had placed it in the context of a Panzer division, a combined arms organization built around the tank and including artillery, engineers, and infantry.[34] Today's organizational transformation requires that with the shift from "mass destruction" to "precision warfare" comes a parallel shift from mass armies to smaller, more highly educated, and capital-intensive professional armed forces whose units are commanded by a more decentralised decision-making structure and can be specifically tailored to the task at hand.

The emphasis on high-quality weaponry has reduced the relative importance of numbers and placed a premium on high-quality troops. As a result, the new military will rest primarily on professional forces as the balance between quantity and quality shifts in favour of quality. "At long last, after a reign of almost two centuries, the age of the mass military manned by short-service conscripts and equipped with the products of high-volume military manufacturing is coming to an end."[35]

At the same time, the centralised decision making of the bureaucratic organization, which dominated the industrial age, must change to reflect the decentralization of the information age. Advanced military technologies will, in theory, allow soldiers to know as much about a battlefield as generals. It follows that military organizations must become "de-layered," with soldiers accorded more leeway in taking initiatives. Local command is likely to be empowered at the expense of theatre-level command.

Such a change must be accompanied by new command protocols. While soldiers today can request firepower, these requests must be routed through the chain of command. On a digitized battlefield this process would be largely circumvented. By directly linking soldiers to weapon systems and operators, the digitized battlefied would enable any soldier to give orders as well as receive them. Thus, technically the lower ranks would have the capability to steer the course of battle. "Chaos is a certain by-product of the innovations to come, if there is not a thorough understanding and control over who directs firepower and support assets during a battle."[36] Therefore, a new balance will have to be struck between those who lead and those who follow.

As the organizations change, so too must the people, with new career possibilities, educational requirements, and promotion paths becoming essential. Most notably, increased responsibility at lower echelons must be coupled with enhanced skill levels. "At the very least, flattening the Army's warfighting structure will necessitate a radical revision of current programs for educating and training leaders, especially at the operational level."[37] Moreover, the changing nature of warfare will call for new types of skills. Already, for example, U.S. Navy leaders are finding that the service needs more sailors and officers skilled in information operations, including maintaining computer networks and troubleshooting.[38]

Finally, new military organizations must have the flexibility to switch rapidly from one type of contingency to another. This is best achieved through "flexible force packaging," a concept that demands that units from one organization be interoperable with those of another. An armed force whose components are designed with a view toward synergy with one another is one that can best adapt to a given contingency, whether it be an intrastate ethnic conflict, a regional hegemonic threat, or a confrontation between great powers and their allies.

CONCLUSION

Thus, the changing nature of warfare today is characterised by the dramatic technological advances of the past two to three decades, which in turn are demanding fundamentally new military doctrines and organizations. Some of these changes are already in place, while others represent possible future developments, to be implemented in the years and decades to come. As dramatic as they are and potentially will be, it is important to remember that the notion of revolutionary change in warfare is not unique to today. Rather, it is a phenomenon that has appeared repeatedly over the course of history as technological innovations have provided nations with new opportunities to gain the upper hand in warfare. To better understand the current RMA, it is important to take a step back and look at the nature and significance of major military transformations that have influenced the course of history over the past several centuries.

CHAPTER TWO

The RMA in Historical Perspective

"If you don't know history, you don't know much," a prominent U.S. defence expert once told a class of graduate students, and this is certainly an appropriate dictum here.[1] Indeed, it is impossible to fully understand the revolution in military affairs without placing it into some sort of historical perspective. Doing so involves not only looking at the origins of the current RMA but also having a general sense of the course and outcome of previous RMAs. Military historians and strategic thinkers have offered various ways of looking at the nature of military revolutions in history and, in doing so, have offered valuable insights into the current military transformation. What follows here is a discussion of the theses advanced by a select few of these thinkers, in order to illustrate the historical dimension. From this analysis it is possible to draw out some general points about revolutions in military affairs, before turning to the origins of the current RMA, its strategic drivers, and some cautionary points about the promises it offers.

PERSPECTIVES ON RMAS AND MILITARY REVOLUTIONS

Alvin Toffler and Heidi Toffler

Perhaps two of the best-known thinkers in this area are the futurists Alvin Toffler and Heidi Toffler. In their book *War and Anti-War*, the Tofflers offer a sweeping historical view of military transformation.[2] They begin by noting that the term "revolution" is often applied rather loosely to technological changes, such as the introduction of gunpowder, the airplane, or the submarine. Even though these inventions had a significant impact on history, they were really "subrevolutions" in that they basically created better or more efficient ways of doing things within the existing "game." A true revolution, they argue, "goes beyond that to change the game itself, including its rules, equipment, the size and organization of the 'teams,' their training, doctrine, tactics, and just about everything else." In this sense, the Tofflers' understanding of an RMA echoes that of the U.S. Office of Net Assessment in that it must encompass not only technological changes but also doctrinal and organizational developments. But the Tofflers also go a step further by arguing that a true military revolution "changes the relationship of the game to society itself." By this measure, military revolutions have occurred only twice before in history, and, most likely, a third such revolution is currently underway.

The Toffler's central thesis is that throughout history the way men and women make war has reflected the way they make wealth. How people work and make wealth, in turn, has undergone two "waves" of change and is now in the midst of a third wave. "First-wave" premodern societies were centred on agriculture, and most people made their livelihood toiling by hand. War reflected this form of livelihood in that most wars were over the control of agriculture, armies were paid for and raised by landowners, and warfare was carried out primarily through hand-to-hand combat.

The industrial revolution launched the second wave of historical change, transforming the way early-modern societies made a living. Once more, the manner in which wealth was created came to be reflected in warfare. Just as mass production became the core principle of industrial economies, so too did mass destruction become the core principle of industrial-age warfare. Mass production corresponded in military affairs to the *lévee en masse* – the conscription of mass armies paid by and loyal to the modern nation-state. The idea of a nation in arms, in turn, was a product of the French Revolution, which roughly marked the transition between the old agrarian regime and the rise of an industrial state.

Other parallel changes could be seen. The bureaucracies of the new industrial economy led to the creation of bureaucratic general staffs in the military. The standardization of industrial output was soon applied to military weapons, training, organization, and doctrine as the ragtag armies of the agricultural era were replaced with standing armies trained in military academies. Whereas the primary implements of war in the old regime were bayonets, swords, and arrows propelled under human force, second-wave tools of warfare increasingly reflected industrial age machinery. The machine age gave birth to such things as the machine gun, mechanised warfare, and the strategic bomber. The mass production of industry was increasingly reflected in the mass destruction of warfare as the trend towards "total war" or "absolute war" reached its epitome in the development of the atomic bomb.

The Tofflers argue that in the 1970s and early 1980s, a new, third-wave economy began to take hold, driven not by land, labour, raw materials, and capital but by knowledge, data, and information. The new economy is characterised by the "demassification" of mass production, offering a much wider variety and an often-customised range of products. The kinds of people who make these products are significantly different from the workers of yesterday, with the low-skilled, interchangeable muscle work of the second wave economy being replaced by more specialised skill requirements. At the same time, small, differentiated work teams have replaced the vast groups of workers doing the same work. Large, bureaucratic, second wave organizations have re-engineered themselves into smaller, more flexible companies.

For the Tofflers, the Gulf War of 1991 was the first to reflect the new, third wave way of wealth creation. In this "dual war," one war was fought with second-wave weapons designed to create mass destruction, while another war was fought with third-wave "smart" weapons designed for pinpoint accuracy, customised destruction, and minimal "collateral damage." Like the economy, warfare was becoming demassified, with the goal being ever more precision and selectivity. Just as knowledge had become the centre of economic activity, knowledge was also becoming the crucial element in warfare. And just as the "smart" economy increasingly required smart workers, smart weapons needed smart, more highly skilled soldiers. Finally, echoing developments in the civilian world, military organizations began to be reorganised into smaller, more flexible formations. Military leaders came to realise that fewer people with intelligent technology could accomplish more than many people with the brute-force tools of the past.

Andrew Krepinevich

Andrew Krepinevich,[3] one of today's foremost thinkers on the current revolution in military affairs, has also offered a framework for looking at military revolutions over the course of history.[4] For Krepinevich, military revolutions comprise four elements: technological change, systems development, operational innovation, and organizational adaptation. By these criteria, he argues, there have been as many as ten military revolutions since the fourteenth century. The infantry revolution of the Hundred Years War saw infantry displace cavalry as the dominant combat unit on the battlefield. It was fuelled by the technological development of the longbow and the doctrinal innovation of integrating archers with dismounted men-at-arms. This revolution was followed by the artillery revolution, in which technological improvements in cannons increased the power of artillery fire and, within decades, reversed the centuries-old dominance of the defence in siege warfare. No longer able to rely on castles for protection, defenders had to move the contest into the field – with all the accompanying organizational and operational changes that that entailed. The fortress revolution of the sixteenth century effected a kind of comeback against the artillery revolution with a new style of defensive fortification that used thicker walls and could better withstand artillery bombardment. But because these fortresses were so expensive to construct, the focus soon shifted back to infantry and the field. The gunpowder revolution was driven by the technological innovation of musket fire and the tactical innovation that saw infantrymen abandon the tight squares of the past in favour of drawing up in a series of straight lines. This allowed for a continuous stream of fire as one rank fired while the other reloaded. Meanwhile, the revolution of sail and shot caused the character of conflict at sea to change dramatically as European navies moved from oar-driven galleys to sailing ships that could exploit the artillery revolution by mounting large guns.

Krepinevich argues that the next dramatic change in warfare was the napoleonic revolution, in which the introduction of the *lévee en masse* following the French Revolution brought about a quantum leap in the size of armies. At the same time, the Industrial Revolution allowed the French to standardise their artillery calibres and carriages and fabricate interchangeable parts. The land warfare revolution followed in the mid-nineteenth century, propelled by the civilian technological advances of the railroad and telegraph. These new technologies allowed military leaders to sustain large armies in the field, mass forces quickly at a specific location, and coordinate widely dispersed operations. The late nineteenth and early twentieth centuries saw a naval revolution during which wooden sailing ships gave way to metal-hulled ships powered by turbine engines, thereby transforming the character of war at sea once more. The mature phase of this revolution found Britain trying to sustain its naval position against Imperial Germany in the lead-up to World War I. The latter stages of the war set in train a number of technological advances that led to the interwar revolutions in mechanization, aviation and information and ultimately set the stage for the character of warfare in World War II, including combined blitzkrieg land/air operations, carrier aviation, amphibious warfare, and strategic aerial bombardment. Finally, the nuclear revolution of the mid-twentieth century centred on warfighting doctrines and organizations that called for avoiding war altogether.

Williamson Murray

A third thinker is Williamson Murray,[5] who has also suggested how one might think about RMAs of the past and the implications of the historical record for the future.[6] Making a distinction between military revolutions and revolutions in military affairs (and echoing the views of the Tofflers), he argues that military revolutions not only fundamentally change the character of warfare but also recast the nature of society and the state. By these criteria, Murray argues, there appear to have been four military revolutions in recent centuries. The creation of the modern nation-state was based on organised and disciplined military power in the seventeenth century. The French Revolution established new norms for the mobilization of economic and popular resources in wartime. The Industrial Revolution was first felt in warfare during the American Civil War when the railroad, steamboat, rifled musket, artillery, and telegraph were exploited to their full advantage. Finally, World War I not only set the seeds for combined arms, strategic bombing, submarine warfare, carrier operations, and amphibious warfare but also sparked immense political, economic, and social changes, including the rise of communism and the end of empires.

For Murray, RMAS come after the larger phenomenon of military revolutions: "RMAS involve putting together the complex pieces of tactical, societal, political, organizational or even technological changes [brought about by the military revolution] into a new conceptual approach to war." Thus the RMA that followed the creation of the nation state encompassed Dutch and Swedish tactical reforms. The French Revolution led to the Napoleonic way of war. The Industrial Revolution presaged the RMA that centred on the railroad. And World War I spawned numerous RMAS, including combined arms, strategic bombing, carrier operations, and submarine warfare. The question today is whether the political, economic, and societal changes brought about by the ongoing information revolution will prove to be a military revolution and bring about the third-wave revolution in military affairs predicted by the Tofflers.

Clifford Rogers

Like Murray, Clifford Rogers[7] also makes a distinction between RMAS and military revolutions.[8] In contrast to Murray (and the Tofflers), however, Rogers sees RMAS as preceeding, rather than following, military revolutions. Such revolutions take place when an RMA has wide-ranging implications for social, economic, and political structures, balances of power, and other areas outside the realm of the armed forces. Rogers gives as an example the artillery revolution, which reversed the tactical advantage from defence to offence because newly vulnerable castle walls no longer allowed weak armies to resist strong ones. This change brought great advantages to the few powers with the resources to maintain large, powerful armies and eventually led to the emergence of the centrally governed nation-state with revenues and a standing army.

Rogers highlights some of the questions that need to be posed if one is to identify whether an RMA is likely to become a full military revolution: Does the current RMA involve a military change in direction or is it "the same only more so"? Does it change the balance between the offence and the defence? Does it change the balance between small states and large populous ones? Does it change the types of components that must be considered in order to assess a nation's military strength? To what extent will it require changes in social, cultural, or economic structures, as opposed to just military ones? And does it mean a major difference in the answer to the question of who fights in society?

Rogers argues that based on these indicators, the current RMA is indeed likely to become a full military revolution. The new style of warfare seen during the Gulf War does not seem to be a case of the same only more so. If, as airpower proponents argue, airpower surpasses land forces as *the* decisive element in war, it will represent a clear change in direction in the nature of war. At the same time the new technologies employed during the Gulf War imply a major change between the offence and the defence. Advances in speed, mobility, and surprise are likely to reduce the effectiveness of the defence and place the advantage with the offence. The current RMA is also likely to change the fundamental components of power. Whereas previously the most important indicators of a nation's warfighting potential were its population as a whole and the strength of its overall economy, increasingly it is the quality of its standing forces and the high-tech sector that will be the determining factors. This opens up the possibility for smaller states to play in the military major leagues in a way that they generally have not been able to do since the French Revolution. Finally, the current revolution indicates a major change in the answer to the question of who fights? In contrast to the large conscript armies of the past, the new style of warfare emphasises the role of a professional armed force made up of troops with highly technical and specific skills.

GENERALLY SPEAKING

Taking these thinkers together, it is possible to draw out some general points about revolutions in military affairs. Perhaps most importantly, RMAs take more than forward leaps in technology. While advances in technology normally underwrite a military revolution, they must be incorporated into new doctrines and executed by new organizational structures before a revolution is possible. In addition, there is no common transition period from one military regime to another; rather, the rate of transition is typically a function of the level of competition among major players in the international system.[9] Most RMAs take considerable time to develop, even in wartime, and peacetime RMAs can take decades.[10] A third point is that RMAs normally bestow an enormous and immediate military advantage on the first nation to exploit them in combat.[11] But the competitive advantages of a military revolution are usually short-lived. Challengers typically recognise the potentially great penalties for not keeping up and take action to maintain the competitive position of their military organizations.[12]

Technologies that underwrite a military revolution are often originally developed outside the military sector and then exploited for their military applications. In past RMAs, many of the key systems were already in use in the civilian world decades before significant changes occurred in military organizations. Railroads, for example, began to carry commerce in the 1830s but were not used extensively for military operations until the 1860s. A further point is that although the source of revolutions in military affairs is often technological (whether civilian or military), the driving force for change in the conduct of war can also lie in the realm of political or economic life. Perhaps the most vivid example of this was the societal change in France that led to the French Revolution and the ultimate creation of large national armies. Finally, a revolution in military affairs does not necessarily obviate or replace preceding RMAs. As was demonstrated during the Gulf War, older forms of warfare (which in their time were revolutionary) can continue to exist even as new forms take hold.

ORIGINS OF THE CURRENT RMA

These general points provide a useful backdrop for looking more closely at today's revolution in military affairs. The current RMA had its origins in the latter half of the 1970s and in what former U.S. defense secretary Harold Brown called the "offset strategy." At that time, NATO estimated that if it were faced with a surprise armoured attack from the Warsaw Pact the West would be outnumbered three to one in personnel and armoured equipment. The United States therefore developed a strategy of using the West's technological superiority to offset the quantitative advantage held by the Soviets. The offset strategy focused not so much on building better tanks, guns, or aircraft but on supporting them on the battlefield with newly developed systems that would multiply their combat effectiveness. This strategy was pursued consistently by five u.s. administrations in the 1970s and 1980s and centred, in particular, on three military support systems: command, control, communications, and intelligence; defence suppression; and precision guidance.[13]

The Soviet Union followed the developments in the United States closely. In the late 1970s relatively lower ranking Soviet officers began to argue that computers, space surveillance, and long-range missiles were merging into a new level of military technology significant enough to shift the balance of power between East and West.[14] These concerns soon rose to the most senior levels of Soviet military command, prompting Marshall Nikolai Ogarkov, chief of the Soviet General Staff at the time, to write a series of papers on what he called the military technical revolution. Ogarkov's central concern was that by the mid-1980s the United States would have solved its strategic problem "by synthesising new technologies, evolving military systems, operational innovation and organizational adaptation into a whole that was more powerful than the parts."[15] Ogarkov began to champion a sweeping reorganization of the Soviet command and force structure, as well as the incorporation of these new technologies into Soviet conventional forces. But such an overhaul would have required massive investments of capital that the Soviet Union could not afford. Moreover, the political will did not exist for such dramatic change; most politicians perceived the strategic implications of such a plan as entailing a dangerous rivalry with the West. As a result, although a few organizational changes were undertaken, Ogarkov's vision was essentially abandoned.[16]

Many Soviet watchers in the United States dismissed Ogarkov's talk of a military technical revolution as Soviet propaganda. They assumed that the Soviet officers were talking about purported advances in *Soviet* technology rather than Western technology. But Andrew Marshall, in the Office of Net Assessment of the u.s. Department of Defense, was increasingly convinced that the ussr was in decline and that Soviet analysts were actually reacting to the promise of Western technology. The fall of the Berlin Wall in 1989 and the subsequent crumbling of the Eastern Bloc confirmed Marshall's premonition and prompted him and his colleagues to begin thinking about an alternative paradigm to the bipolar world.

Marshall suggested that an historical analogy may be the interwar period, when numerous technological advances and doctrines were developed that ultimately played a revolutionary role in World War II. They noted that in this case, as in other revolutionary periods, the big changes in military capability had occurred when new technologies had been accompanied by shifts in tactics, doctrine, and organization. By 1993 Marshall and his office had declared the Soviet term military technical revolution to be too narrow and argued instead that warfare was in the early stages of a revolution in military affairs.[17]

Meanwhile, the United States continued to exploit developments in microelectronics and computers as a means of offsetting the numerical superiority of Soviet troops in central Europe. During the 1980s the Pentagon conceived, developed, tested, produced, and deployed systems embodying advanced technologies. The U.S. military developed tactics for using the new systems and conducted extensive training exercises with them, mostly under simulated field conditions. But while the effectiveness of the weapons had been demonstrated in many exercises, they had never been used in war. It was not until the Gulf War that U.S. military leaders came to understand the remarkable increase in military capability that the new technologies would provide.[18] Despite the drama of the Gulf War, however, in the early 1990s the U.S. military was content to avoid rapid change. The military, a naturally conservative institution, was not very interested in pushing a revolution in military affairs, and few people in the executive branch or Congress were either. Things probably would have stayed in this prerevolutionary stage had it not been for the arrival in the Pentagon of two high-level RMA champions.

In 1994 U.S. president Bill Clinton appointed a technologically minded secretary of defense. William Perry had served under Harold Brown when the offset strategy was first developed and had already written about the revolutionary nature of the Gulf War. In his view, what distinguished American operations in Desert Storm from previous conflicts was a refined approach to warfare that allowed intelligence sensors, communications, and precision weapons to be linked together. That same year, Admiral William Owens was appointed vice chairman of the Joint Chiefs of Staff and soon became the foremost military advocate of revolutionary change. For

Owens, the technology was already there – a result of the previous two decades of Cold War investments. The separate improvements in the capacity to collect information about a battlefield, assign targets to the right forces, and destroy those targets with precision from longer ranges were all individually very impressive. But the integration of these capabilities had not, in Owen's view, been taken far enough. Owens argued that if the new advanced surveillance, command and control, and precision force subsystems could be integrated into a system of systems the United States would be propelled into a "qualitatively new order of military power."[19]

Although the systems integration that Owens championed carried substantial implications for operational and force structure, he never articulated what they were. This was left to the chairman of the Joint Chiefs of Staff at the time, General John Shalikashvili, who took a first cut at such a task in his vision statement of 1996, *Joint Vision 2010*. This document has been described as "a sophisticated exposition of how the technological integration posed by [Admiral] Owens' system-of-systems can and should be used in military operations, and an initial discussion of the kind of force structure needed to conduct such operations."[20] *Joint Vision 2010*, now updated with *Joint Vision 2020*, remains the overall conceptual template for the future of America's armed forces.

THE RMA AND AMERICA'S POST-COLD WAR STRATEGIC POLICY

The idea of an RMA gained currency at a critical juncture in America's post–Cold War strategic thinking. Contrary to those who have viewed U.S. security policy after the Cold War as incoherent or lacking direction, Michael Mastanduno has argued that the administrations of both George Bush Sr and Bill Clinton followed a consistent strategy of pursuing the preservation of America's preeminent global position.[21] In an effort to prolong the "unipolar moment," the United States adopted policies of reassurance towards the status quo states of Germany and Japan, to convince them to remain partial great powers; of integration and engagement towards the undecided states of China and Russia, to encourage them to take part in a U.S.-centred international order; and of the pursuit of multilateral processes in its other foreign policy undertakings. In this context, the United States took on the responsibility for addressing threats not only to its own interests but also to those of its allies and friends and indeed any threats that could seriously unsettle international relations.[22]

Such a role does not come cheaply either in monetary terms or political terms. The great issue of contemporary U.S. foreign policy, one analyst has noted, is the "contradiction between the persisting desire to remain the premier global power and an ever deepening aversion to bear[ing] the costs of this position."[23] It is in response to these conflicting desires that the RMA becomes strategically attractive. Monetarily, the revolution in military affairs offers the United States the possibility of doing "more with less," enabling it to maintain its global military power even at a time of shrinking U.S. defence budgets. And politically the RMA responds to the imperative first highlighted during the Gulf War and later reinforced in the Kosovo campaign of conducting rapid, surgical, almost casualty-free warfare. "Whether true or not, many decision makers now believe that the American people will only support the use of force abroad if it promises smashing victories with few or no casualties."[24] Edward Luttwak has elaborated the view that in future only those forces that are least exposed to casualties, such as high-technology stand-off forces, will be "usable" in a domestic political context.[25]

SOME CAUTIONARY NOTES

Although the RMA presents many opportunities, there are good reasons to view some of its promises with caution. In the first instance, some elements of the RMA vision may be unachievable. Michael O'Hanlon has taken to task, for example, the air force assertion in its most recent service statement that "With advanced integrated aerospace capabilities, networked into a system of systems, we'll provide the ability to find, fix, assess, track, target and engage anything of military significance, anywhere." This may be true for easily recognisable military objects like tanks and ships, but it is less true for antiaircraft and antitank missiles carried in trucks, enemy soldiers interspersed among civilians or hiding in buildings, or weapons of mass destruction buried underground. Optical and infrared sensors cannot see through bad weather, nor can they see through most solid objects. Radar and radio waves can penetrate clouds and rain, but they cannot penetrate metal containers and are quickly stopped by soil and water. These laws of physics will be in place "not just in the near term, but indefinitely."[26]

Second, having more advanced technology than one's opponent does not constitute an automatic formula for winning a war. Many RMA proponents focus on the merits of "dominant battlespace awareness," or the ability to see all enemy and friendly forces and platforms in the theatre of war, and "dominant battlespace control," or the ability to receive and transmit information and decisions in real time. But increased information, or data, does not equate with increased knowledge and understanding – indeed, it could just as likely lead to sensory overload.[27] Similarly, "while the RMA greatly increases the speed of data transmission, it has not appreciably quickened political and strategic decision-making or improved its quality."[28] In short, advanced technology can go a long way towards decreasing the fog and friction of war, but it cannot eliminate them altogether.

An excessive focus on advanced technologies could cause military leaders to ignore or place less emphasis on other key requirements for winning wars. "The problem with the extreme view of the RMA is not the attempt to leverage technology in the pursuit of revolutionary change, but in *technocratic thinking*: the belief that an edge in technology itself is enough. Technocratic thinking can lead to a dangerous de-emphasis of other factors critical to success in war."[29] These things include readiness, force structure, and, perhaps most importantly, training. Stephen Biddle has demonstrated that the main cause of the one-sided coalition victory in the Gulf War was not technology per se but the skill differential between coalition and Iraqi forces.[30]

The promises of the RMA could also lull leaders into planning only for short, rapid victory when a longer engagement may in fact be necessary. As Stephen Blank has emphasised, the first operation becomes the only operation only if it succeeds; if not, leaders are back to essentially the same drawing board that they faced before the advent of long-range precision strike munitions. But for the political and military success of a mission it is best to plan for such an eventuality in advance, rather than reacting to events as they happen. "As we now know there was no 'plan B' for Kosovo. NATO had to scramble to come up with one, and that option resembled an old-fashioned war of attrition."[31]

America's commitment to one vision of future warfare could blind it to the emergence of radically different forms of warfare. Lonnie Henley has highlighted revolutionary developments in miniaturization, nanotechnology, and biotechnology and argued that the synergistic merger of these trends could, in perhaps as little as ten to fifteen years, bring about "a technological transformation even more profound than the information revolution that is the focus of current attention."[32] Unlike the *Joint Vision 2010* and *2020* form of warfare, however, these are areas where the United States has not made significant strides in applying new operational concepts to emerging technologies. This, in turn, has left the door open for other nations to "strike out in a direction radically different from the prevailing paradigm of military operations, and develop capabilities that their opponents are unable to counter."[33]

Finally, pursuing a more extreme version of the RMA could be considered akin to placing all of one's eggs in one basket and therefore putting at risk the ability to effectively manage and respond to a future crisis. As one analyst has commented, what if the RMA proponents are wrong? What if the United States gets rid of all its heavy tanks and the next war really does end up looking like the last war?[34] In this sense it seems clear that radical innovation is not always good. If the wrong ideas are adopted, transforming a force or its tactics could even cause harm.[35] The next chapter examines the efforts at force transformation that the U.S. military is taking to harness the technological, doctrinal, and organizational tenets of the RMA.

CHAPTER THREE

Is There an RMA? An Assessment of U.S. Force Transformation

Many RMA observers are quick to point out that the fairly short time frame implied by the term "revolution" – strictly speaking a "complete change, turning upside down, great reversal of conditions" – has not been met. They note that several of the technologies that are considered today to be revolutionary – technologies such as precision munitions, reconnaissance satellites, infra-red vision devices, laser range finders, cruise missiles, the global positioning system, and stealth materials – first made their debut in the Cold War era.[1] Moreover, many of the operational concepts that are part of the RMA – like air-ground co-ordination, three-dimensional naval combat, and tactical command and control communications – merely represent the full maturation of ideas that first emerged in the 1930s and 1940s.[2] It follows that what we are really experiencing are not so much revolutionary as evolutionary changes in the nature of military operations.

But this fact does not obviate the potential for revolution. Rather, it only illustrates that "the current *revolution,* like all others, is built on a long foundation of *evolutionary* change."[3] As Andrew Marshall has stated, "the term 'revolution' is not meant to insist that the change will be rapid, but only that the change will be profound, that the new methods of warfare will be far more powerful than the old."[4] Those who ascribe to the evolutionary view argue that the RMA is best understood as a *process of transformation.*[5] A recent study commissioned by the chairman of the joint chiefs of staff makes this link directly: "Transformation refers to the set of activities by which DOD attempts to harness the revolution in military affairs to make fundamental changes in technology, operational concepts and doctrine, and organizational structure."[6] Some experts argue that it is far from obvious that military technology is now poised to advance even more quickly than it has in the last half century.[7] Others take the position that within the last decade or so evolutionary changes have reached what might be called potential "take off" velocity.[8]

This chapter examines the extent to which the United States can be expected to transform its military over the next two to three decades. It begins at the conceptual level, looking at the degree to which key technological, doctrinal, and organizational tenets of the RMA are reflected in current U.S. defence policy and service visions. It then moves on to examine in practical terms the concrete steps that the DOD and the U.S. military services are taking to exploit the perceived benefits of the RMA. The chapter follows with a discussion of possible barriers to force transformation, before concluding with an assessment of future prospects for change and whether or not a revolution in military affairs will actually be achieved.

U.S. DEFENCE POLICY AND THE VISION THING

One does not need to look far to find references to the RMA and force transformation in official U.S. defence policy. The Quadrennial Defense Review (QDR) of 1997 listed "exploiting the Revolution in Military Affairs" as one of four avenues[9] by which the DOD would seek to "Prepare Now for an Uncertain Future" – the third pillar of America's defence strategy.[10] It committed the DOD to "harnessing new technologies to give U.S. forces greater military capabilities through advanced concepts, doctrine and organizations," the exact recipe for achieving an RMA force, and it directly linked the RMA to the ongoing transformation of U.S. military capabilities.[11] Placing the revolution in military affairs first on its list of key military-technical trends, the QDR of 2001 also links exploiting the RMA to transforming the country's military forces. Moreover, it states that "transforming defense" is at the heart of America's strategic approach to achieving its defence policy goals.[12]

Joint Vision

The conceptual basis for U.S. force transformation is found in each of the service visions, as well as in the joint vision that is put out by the chairman of the joint chiefs of staff. *Joint Vision 2010*, released in 1996, was a landmark document, in that it was the first U.S. vision statement to put into a coherent whole a myriad of technological, doctrinal, and organizational changes associated with the RMA. Without ever mentioning the term "revolution in military affairs," *Joint Vision 2010* discusses technological advances in sensor, command and control, precision force, and stealth capabilities; doctrinal changes towards increased jointness, littoral warfare, rapidly mobile yet highly lethal ground forces, and long-range precision air power; and organizational changes towards smaller, more tailorable units, decentralised command and control, and a reduced number of launch platforms and a reduction in the ordnance required to destroy targets. *Joint Vision 2010* centres on the achievement of four new operational concepts – dominant manoeuvre, precision engagement, full-dimensional protection, and focused logistics – all of which incorporate RMA-related technologies, doctrines, and organizational changes. The overall objective is to achieve "full-spectrum dominance," that is, the ability to dominate an opponent across the range of military operations.

Joint Vision 2020, released in May 2000, carries forward the ideas expressed in *Joint Vision 2010*, stressing the need to harness RMA technologies in order to turn today's manoeuvre, strike, force protection, and logistics capabilities into tomorrow's new operational concepts. It restates the thesis that full-spectrum dominance is the overarching objective for a transformed U.S. military and challenges each of the services to field forces that are faster, more agile, more precise, better protected, more rapidly deployable, and easier to sustain. In contrast to the more technological focus of *Joint Vision 2010*, however, *Joint Vision 2020* emphasises the "need to broaden our focus beyond technology and capture the importance of conceptual innovation as well." When it comes to achieving full-spectrum dominance, the document stresses, even more important than new technology and equipment is the development of doctrine, organizations, training and education, leaders, and people that effectively take advantage of the technology. This is a significant shift in emphasis, considering that an RMA cannot be achieved until technological advances are translated into changes in doctrine, organization, and force structure.

The new vision statement also offers a subtle, yet sober, second look at the RMA. For example, although *Joint Vision 2010* states that new sensor technologies will not eliminate the fog of war, it places its weight on the view that they will make the battlespace considerably more transparent. By contrast, *Joint Vision 2020* emphasises the opposite nuance: "We must remember that information superiority neither equates to perfect information, nor does it mean the elimination of the fog of war. Information systems, processes, and operations add their own sources of friction and fog to the operational environment." Similarly, while *Joint Vision 2010* indicates that the new form of warfare will be "more efficient in protecting lives," *Joint Vision 2020* is much more cautionary concerning the ability to achieve near casualty-free war: "Achieving full spectrum dominance does not mean that we will win without cost or difficulty. Conflict results in casualties despite our best efforts to minimise them. We should not expect war in the future to be either easy or bloodless." Finally, whereas *Joint Vision 2010* lauds the merits of information superiority during a military operation, *Joint Vision 2020* notes that such superiority is only useful when it is translated into superior knowledge and superior decision making.[13] "Decision superiority does not automatically result from information superiority. Relevant training and experience are also necessary." In short, good technology cannot substitute for good people.

Service Visions

Each of the vision statements of the U.S. military services also contains elements that are central to the RMA. The U.S. Navy's ... *From the Sea*, released in 1992 and updated as *Forward ... From the Sea* in 1994, marked a landmark shift in the navy's focus from addressing a global maritime threat to power projection in the littoral regions in support of army and air force operations. The vision statement also emphasises that projecting power and influence ashore requires naval forces shaped for joint operations. As elaborated in its *Navy Operational Concept* of 1997, the Navy plans to carry out these two RMA-related doctrines – littoral warfare and joint operations – with the application of cutting-edge RMA technologies. Here the focus is on "network-centric warfare," that is, the use of advanced communications systems to pass sensor information rapidly among naval platforms to increase the battle group's ability as a whole to respond to threats. Network-centric warfare gives primacy to sensors and information networks over individual ships or submarines. Eventually, for example, it will likely be possible to pass sensor data so quickly that one warship will be able to shoot down an incoming missile by using another ship's radar picture. With network-centric warfare the force-multiplier effect will lie not so much in the capabilities of specific platforms as in their ability to rapidly communicate with one another.

In many ways, the u.s. Marine Corps can be considered a natural RMA force, in that it has traditionally emphasised two doctrinal elements that are now closely associated with the RMA – expeditionary warfare and joint operations. In *Marine Corps Strategy 21*, released in 2000 the service continues in this tradition, listing several core competencies that echo RMA doctrinal and organizational trends. They include an expeditionary culture, task-organised forces, joint competency, combined-arms operations and forcible entry from the sea. The corps' expeditionary culture is manifested in its Marine Expeditionary Forces (for major theatre war) and Marine Expeditionary Brigades (for smaller-scale contingencies), both of which rely on task-tailored units. The vision notes that the Marine Corps is ideally suited for joint, allied, and coalition warfare, and that it is the only u.s. service specifically tasked by Congress to act as an integrated combined-arms force operating in three dimensions – air, land, and sea. To this end, the corps operates marine air-ground task forces, which, as their title suggests, integrate marine, ground, and air units under a single commander. Finally, it goes without saying that the Marine Corps conducts missions, as it always has, in a littoral environment with a focus on force projection onto shore.

In October 1999 the u.s. Army issued a dramatically new vision for the future. Stung by its inability to deploy rapidly into theatre during Operation Allied Force in and around Kosovo in the spring of 1999, the army undertook to focus on one of the RMA's key doctrinal tenets – the development of a flexible and rapidly mobile, yet still highly lethal, ground force. *Army Vision 2010*, released in 1996 shortly after *Joint Vision 2010*, had focused on the land components of dominant manoeuvre, precision engagement, full-dimensional protection, and focused logistics. Although the vision's dominant manoeuvre section stated the need to be able to commence transport of a versatile, tailorable, modular army within hours of a decision to deploy, it was not specific on how this was to be done.

By contrast, the army's new vision statement, *Soldiers on Point for the Nation: Persuasive in Peace, Invincible in War*, provides a concrete road map for the future. The army's plan is to develop brigade combat teams that can deploy anywhere in the world in four days, followed by the remainder of the division twenty-four hours later, and to be able to deploy five divisions anywhere in thirty days. The overall objective is a rapidly deployable medium-weight force that does not surrender capability in force protection or firepower. The army's vision, in short, is the creation of a force "that combines the decisive warfighting lethality of today's mechanised forces with the strategic responsiveness of today's light forces."[14] Central to this effort will be developing and harnessing relevant RMA technologies.

America's Air Force: Global Vigilance, Reach and Power, released in 2000, is the aerospace answer to the challenges of *Joint Vision 2020*. As in its earlier vision statement of 1996, *Global Engagement*, the U.S. Air Force lists aerospace superiority, information superiority, global attack, precision engagement, rapid global mobility, and agile combat support as its core competencies. The new vision also reiterates that the air force plans to capitalise on the full set of revolutionary technologies – such as stealth, advanced airborne and space-borne sensors, and highly precise all-weather munitions – to achieve these core competencies. What is notable about *Global Vigilance* is that it takes the technological discussion further, indicating the specific doctrinal and organizational means by which the air force will seek to achieve them. These means are closely associated with the RMA. Doctrinally, the U.S. Air Force has undertaken to become an expeditionary aerospace force that can respond rapidly to regional contingencies anywhere in the world. Organizationally, it plans to do so with tailored capabilities that require fewer airlift resources and support personnel. To this end, the air force has reconstituted itself into one comprised of ten expeditionary aerospace forces. The objective is to be able to deploy one such force anywhere in the world within forty-eight hours and up to five additional ones in fifteen days.

Experimentation

Central to force transformation is the experimentation process. In 1999 U.S. Atlantic Command was redesignated as Joint Forces

Command and was given responsibility for joint war-fighting concept development and experimentation. Its experimentation process is organised around three lines of inquiry. "Enhancement of the current force" centres on the period up to 2010 and looks at improvements to existing forces that are critical prerequisites for future transformation. It has demonstrated, for example, the requirement for joint task force headquarters and joint surveillance and command and control arrangements. "Realizing *Joint Vision 2020*" looks beyond current systems and capabilities to determine what is needed to replace them between 2010 and 2020. Finally, "Transforming the force for the Revolution in Military Affairs" focuses on the period beyond 2020 and considers those concepts and technologies that have the potential to effect a truly revolutionary transformation of the joint force.

Training and Doctrine Command serves as the army's lead agent for force transformation. Its Force XXI process, originally mandated to help implement the army's vision statement of 1996, has now been refocused to develop the concepts, doctrine, and leadership needed to field the force outlined in the new vision. The old Army after Next process, which focused on the period beyond 2010, has been replaced with a new Army Transformation War Game series, specifically tailored to support the new vision. Already these games have demonstrated the key imperative of transformed strategic lift. The army also has eight battle laboratories, or "battlelabs," that focus on space and missile defence, manoeuvre support, mounted manoeuvre battlespace, dismounted battlespace, air manoeuvre, battle command, depth and simultaneous attack, and combat service support.

The Naval War College in Newport, Rhode Island, oversees the navy's concept development and experimentation efforts. It works with the Maritime Battle Centre of the Navy Warfare Development Command to co-ordinate Fleet Battle Experiments, which make up the principal navy experimentation activities and take place while the navy is engaged in training exercises. The primary area of focus for the near- to mid-term is network-centric warfare. Experimentation is also taking place on additional capabilities, such as projecting defence ashore with Theatre Missile Defence and precision land attack deep into enemy territory.

The Marine Corps Warfighting Laboratory, part of the Marine Corps Combat Development Command at Quantico, Virginia, carries out the Marine Corps' experimentation with its new Expeditionary Manoeuvre Warfare concept. The warfighting laboratory has developed a five-year, three-phase experimentation process, called Sea Dragon. Each phase starts with limited objective experiments and culminates in a large-scale Advanced Warfighting Experiment. Capable Warrior, which began in 1999 and is the third phase of the Sea Dragon program begun in 1995, is looking at the Marine Corps' ability to conduct littoral warfare in an extended area of operations. Coalition Warrior, scheduled to begin in 2001 after Capable Warrior has been completed, will examine Marine Corps operations in the context of a future coalition and will address associated interoperability issues. Millennium Warrior and Cyber Warrior will focus respectively on Marine Corps operations in smaller-scale contingencies and against asymmetric threats. Millennium Warrior will likely begin in 2003.

The U.S. Air Force carries out its experimentation efforts with advanced concept technology demonstrations and battlelab research and development programs. Its six battlelabs focus on air expeditionary forces, command and control, force protection, information warfare, space, and unmanned aerial vehicles. Examples of recent successful battlelab initiatives include precision targeting with *Predator* unmanned aerial vehicles (UAVs) and battle imaging by Joint Surveillance Target Attack Radar System aircraft, both of which proved very useful during NATO's Kosovo operation. To support the experimentation process, the air force conducts three series of war games, each tailored to a different time frame. The annual Expeditionary Force Experiment focuses on near-term requirements, such as the exploration of capabilities required to provide agile combat support to expeditionary forces. The Global Engagement war games look at joint air and space power in the time frame 2008 to 2020, while the Aerospace Future Capabilities Wargames test alternative force structures for the environment 2020–2025 and beyond. To date, these latter war games have underscored the imperative of integrating space-based capabilities into air, land, and sea operations.

TRANSFORMATION: A CURRENT ASSESSMENT

Thus, force transformation figures prominently in U.S. defence policy and service visions. The Quadrennial Defense Reviews of both

1997 and 2001 make a direct link between the RMA and force transformation and state their centrality to preparing for the future, and all the military vision statements provide conceptual templates for force transformation that are closely linked to the RMA. Each of the services, as well as Joint Forces Command, has also established a concept development and experimentation process to help make their visions reality. Against this backdrop of visions and intentions, taking stock of the RMA requires an examination in practical terms of the concrete steps the U.S. military services are taking to transform their forces.

Navy Transformation

Unlike the army and air force, which have carried forward the conceptual basis found in their earlier vision statements by providing specific guidelines and objectives for force transformation, the U.S. Navy has not yet developed a comprehensive transformation strategy to guide its vision and future evolution. Analysts note that the navy has bits and pieces of a strategy in place, most notably its enunciation of the network-centric warfare concept and its shift from open-ocean warfare to supporting the littoral battlefield. But these concepts have not yet been expanded into a complete road map for transforming today's force into that of tomorrow.[15]

The navy's acquisition priorities reflect the lack of an overall strategy. In support of its move towards littoral warfare, the navy is pursuing the development of a land-attack destroyer, the DD-21, which is likely to be introduced into the fleet at the end of this decade. This multi mission platform focuses on land-attack roles and will use an array of long-range guns and missiles to influence events ashore. Apart from this new destroyer, however, some of the navy's planned acquisitions are inconsistent with the nature of the new international security environment. Most notably, the service continues to centre its fleet on the aircraft carrier, even though it is likely that these large platforms will be increasingly at risk from land-based cruise and ballistic missiles. The navy operates twelve aircraft carriers and has plans to begin building a new generation of carriers to replace its *Nimitz*-class ships in 2006. At the same time, it is purchasing a minimum of 548 F/A-18E/F carrier-based fighters and is developing its own version of the Joint Strike Fighter. Analysts argue it would make more sense for the navy to focus on a force projection platform such as the *Arsenal* ship – a semisubmersible, stealthy barge armed with hundreds of missiles, few sailors, and no tactical aircraft.[16] The navy cancelled this program in 1997.

Marine Corps Transformation

The Marine Corps' transformation strategy is reflected in its doctrinal framework, *Operational Manoeuvre from the Sea*. Instead of massing forces first to achieve a foothold on shore and then attacking a certain objective, the marines intend to lift relatively small teams to the vicinity of a target that may be hundreds of miles inland. They will then rely on the navy (and its land-attack destroyers) to provide precision fire support.

In that it was already an expeditionary force, the Marine Corps is well advanced in implementing this concept of operations, which is relevant to the RMA. The concrete measures it has taken include the development of the V-22 *Osprey* helicopter for increased battlefield mobility. That said, testing and maintenance problems have placed the future of the *Osprey* in question. Moreover, the corps plans to purchase its own version of the Joint Strike Fighter, and it continues to operate some four hundred *Abrams* tanks. At seventy tons, the *Abrams* is likely to be increasingly ill-suited to the future security environment, which, as a result of reduced access to forward bases, is likely to demand that forces be lighter and more deployable.

Army Transformation

The U.S. Army's plan to dramatically overhaul its force structure encompasses a three-pronged transformation process. First, the army is modernising its heavy "legacy force" so that it is better able to carry out current missions. A key feature is the "digitization" of divisions whereby digital technology is being added to or built into aircraft, tanks, artillery, and individual soldier systems with the intention of providing commanders with an instantaneous and complete picture of the battlefield. Phase two centres on converting a selected number of combat brigades into four-thousand person interim brigade combat teams, which are intended to bridge the gap between traditional heavy and light forces. This "interim force" will use lighter, wheeled-combat vehicles, and thus be more rapidly deployable than the army's heavy formations. Finally, the "objective force" will be based on the deployable brigade combat teams but will use significantly more advanced technologies. The centrepiece will be the Future Combat System, which at twenty tons will be radically lighter than the *Abrams* tank and therefore a much more agile and deployable platform.

The army has made progress in each of these areas. It has already fielded its first digitized division, the Fourth Infantry Division (Mechanized), in Texas and hopes to field its first digitized corps by October 2004. In its budget for 2001 the army asked for and received the necessary initial funding for its transition to a medium-weight force, and this was supplemented with additional funding from Congress in the summer of 2000. The army has also selected an interim combat vehicle, the "off-the-shelf" Light Armoured Vehicle (LAV III) to equip the interim brigades. With these funding and equipment decisions taken, the army is well on its way to meeting its target of having between five and eight interim brigades in place by the end of this decade. For the Future Combat System, potential contractors are to submit proposals to the chief of staff of the army by April 2003. If development and production proceeds as planned, the army could begin fielding its objective force as early as 2012.

The army has made some important changes in its procurement priorities to support its transformation efforts. Most notably, in its budget for fiscal year 2001 it scaled back or cancelled six equipment programs that are more suited to a heavier force structure, including a two-thirds reduction in the number of the 110-ton *Crusader* self-propelled howitzer systems it had originally planned to buy. Selecting the off-the-shelf *LAV III* was also important, because its delivery will allow the army to begin reducing the number of its *Abrams* tanks. And fielding the Future Combat System will enable the army to phase out the *Abrams* altogether. Nonetheless, RMA proponents argue that the army has not gone far enough. Despite the reduction in the *Crusader* program, for example, the army still plans to buy some 480 of these heavy howitzers, "even though it could hardly hope to deploy them in a timely or ample fashion to the obscure parts of the world where [the Army] will actually have to go."[17] At the same time, the army continues to devote significant funds to upgrading its heavy (and not very rapidly mobile) MIAI *Abrams* tanks.

Beyond these concerns, critics contend that even with its middle-weight force the army has not fully addressed the nature of the future security environment. Not only must it develop the capability to project substantial land power rapidly and to sustain it indefinitely, they argue, but it must be able to do so in the absence of access to forward bases and large, fixed logistics centres. In addition, the army needs to exploit its potential to conduct precision strikes at extended ranges. "Such deep-strike formations, centered on extended-range reconnaissance (e.g. UAVs and sensors) and strike (e.g., missile artillery and attack helicopters) elements, may represent for this military revolution what the Panzer Division did for Blitzkrieg."[18]

Air Force Transformation

The U.S. Air Force is well advanced in the organizational aspects of its transformation efforts. Plagued by readiness problems and a high pace of operations and taking lessons from the Kosovo operation, in October 1999 the air force undertook to create a "lighter, leaner and more lethal force" with the ability to deploy anywhere in the world.[19] To this end, in 2000 the air force divided almost its entire force – active, reserve, and Air National Guard – into ten Aerospace Expeditionary Forces. Each force comprises 175 aircraft and between ten and fifteen thousand personnel, and each represents a complete aerospace capability, including air superiority fighters, air-to-ground strike fighters with precision weapons, strategic and tactical airlift aircraft, refuelling tankers, bombers, and aircraft to suppress enemy air defences. Each Air Expeditionary Force integrates groups and squadrons that, although geographically separated, are required to train together on a regular basis – a significant step forward from the previous ad hoc deployment procedure, which had the effect of many air force personnel never working together before an operation. With the new organization in place, the air force is now working on increasing its response time to the levels outlined in its new vision statement.

But while many of the organizational changes for an RMA-inspired force are in place, defence analysts have questioned the air force's equipment acquisition priorities. Specifically, the air force has placed its emphasis on exploiting technologies that enhance the central role played by piloted aircraft, rather than looking more generally at how to improve its ability to deliver long-range precision firepower.[20] Thus the nonstealthy F-15 air superiority fighter is being replaced by the stealthy F-22, which will be the most advanced aircraft of its kind in the world and will also incorporate an air-to-ground role. Although it has reduced the original plan to purchase 438 F-22s,[21] the service still plans to buy 339 of the aircraft, with the first F-22 squadron set to become operational in 2005.[22] At the same time, the Joint Strike Fighter is being developed for the air force to replace its F-16 ground strike aircraft. The service plans to buy 1,763 of the aircraft, which will begin replacing the oldest F-16s around 2010.

Military trends indicate that tactical aircraft will not be well suited to the nature of the new international security environment. The relatively short range of the F-22 and the air force version of the Joint Strike Fighter make them dependent on overseas bases that may not be available during wartime. The proliferation of nuclear, biological, and chemical weapons and their means of delivery are likely to reduce access to overseas bases. And the Western world's concern with limiting casualties dictates that, depending on the particular crisis, political leaders may not want to allow their fighters to fly below a certain level, lest they be at risk from antiaircraft fire (as was the case in Kosovo).

This combination of factors means that it may make more sense for the air force to develop a range of capabilities that might better be able to carry out many, if not most, of the missions currently performed by tactical air forces. This includes a greater emphasis on acquiring stealthy unmanned combat aerial vehicles (UCAVs) for precision-strike operations. Although the air force has a UCAV under development, it is not scheduled to reach combat units until about 2010. As an interim measure, it has recently begun to arm its *Predator* UAVs with precision munitions. But while the air force has created some UAV squadrons, it has yet to establish an experimental UCAV wing. The air force may also want to reopen its production line of stealthy B-2 bombers. These long-range power projection platforms operated directly from the United States during the Kosovo and Afghanistan campaigns, proving militarily very effective in delivering satellite-guided Joint Direct Attack Munitions against ground targets.

Jointness

Apart from the shortfalls in individual service transformation efforts, it is not clear that there has been concrete progress towards increasing jointness among the services. Although U.S. Atlantic Command has, as mentioned, been renamed U.S. Joint Forces Command, its commander must rely largely on the power of persuasion to fulfil the joint role, because he has no enforcement authority on DOD procurement councils.[23] The commander has a seat on the Joint Requirements Oversight Council (JROC), which was first created by the Goldwater-Nichols legislation in 1986. But although the JROC was given a revised and strengthened mandate in 1995 to promote jointness by competing military programs across service boundaries, in practice it has had little significant effect on the allocation of defence resources.[24] For example, the QDR of 1997 announced a reduction in the number of JSTARS aircraft, even though these aircraft, operated as they are by the air force and providing a picture of the battlefield for army and marine corps units on the ground and navy ships offshore, strongly promote jointness. Since that time, interoperability among services has been given an increased profile as a "key performance parameter" in the development of individual weapons systems.[25] Finally, with respect to joint experimentation, it is instructive to note that over 90 percent of the U.S. military's experimentation continues to be carried out within the individual services.[26]

Thus, an examination of the U.S. military services' concrete transformation efforts leads one to question whether the RMA will come to fruition in the next two to three decades. While today's American military is smaller and more technologically advanced than the one that fought the Gulf War, in the view of many defence experts it is marked by essential continuities in force structure, equipment, and doctrine.[27] As one analyst has put it in assessing the current state of the RMA, although the U.S. military has a lot of new technology, when it comes to new doctrine and force structure it is "a long way from this, a very long way."[28]

Moreover, despite the transformation rhetoric in the QDR of 1997 and in the recent service vision statements, tomorrow's military is projected to look much the same as today's. "A Navy that buys Joint Strike Fighters and aircraft carriers will find itself operating in

50 years in fundamentally the same manner as it does today. Likewise, current Pentagon plans call for the air force to be primarily comprised of short-range tactical aircraft and most of the army to be driving in land combat vehicles at a top speed of 50 miles per hour."[29] In this context, it is not surprising that the National Defense Panel report that paralleled the QDR in 1997 argued that while the service visions contain many of the capabilities associated with a revolution in military affairs, the procurement priorities of the services do not fully reflect their visions.

WHY THE SLOW PACE OF TRANSFORMATION?

The QDR Framework

A number of factors explain the slow pace of U.S. military transformation. In the first instance, the QDR of 1997 did not provide a framework that promoted rapid force transformation. Its drafters examined three possible courses of action – focus on near-term demands, prepare for a more distant threat, and balance current demands with an uncertain future – and decided on the third. In practice this meant that while the QDR espoused information superiority, dominant manoeuvre, precision engagement, and full-dimensional protection, it also explicitly decided not to pursue these goals too fast.

Part of the reason was budgetary. The QDR of 1997 has been criticised as having been "budget-driven" rather than "strategy-driven"; that is to say, it laid out what the Pentagon could afford, rather than what U.S. national security interests required. According to the QDR of 1997, preparing for a more distant threat – the emergence of a peer or regional competitor in the 2010–2015 time frame, as well as increased asymmetric threats – would have required an investment of at least $100 billion per year, with $65 billion devoted to procurement. By contrast, the path the Pentagon chose had somewhat lower and more manageable figures, at $90 billion and $60 billion respectively.

Vested interests were also undoubtably part of the equation. The dominant military service cultures continue to be centred on armoured warfare on land, tactical fighters in the air, and carrier battle groups at sea. As a result, "there is less enthusiasm in the

Pentagon for transformation than you might believe from listening to statements made for public consumption."[30] The path the QDR of 1997 chose included keeping the Pentagon's two-war strategy in place, and this in turn justified the maintenance of a somewhat smaller, yet essentially unchanged, navy, army, and air force structure and accompanying equipment. The strategy called for the United States to be able to respond to two major theatre wars in overlapping time frames and required almost the same base force structure as did the Bush administration's Base Force of 1990, and the Clinton administration's Bottom Up Review of 1993.

Nor was Congress overly enthusiastic about making the hard choices necessary to move more rapidly towards harnessing the RMA's potential for the U.S. forces. The DOD is saddled with numerous military bases and other defence installations that the Pentagon would prefer to have closed. The funds spent on unwanted infrastructure could be spent on new technologies and equipment. However, it is politically very difficult for any individual Congressman to agree to a base closure in his or her electoral district, because of the numerous jobs at stake. Similarly, it is politically difficult for Congress to shift investments to future weapons systems and away from systems already providing jobs and revenue in the hard-pressed defence industrial sector. Shipyards, for example, can provide hundreds or thousands of jobs in a district.

The combination of these factors meant that even as it accepted the RMA hypothesis, the DOD made few plans to reorganise its main combat units, increase their jointness, alter priorities within the weapons modernization program, or divest itself of unneeded infrastructure. In short, the DOD sought to meet the new strategy of Shape, Respond and Prepare with a force structure essentially organised and equipped for the Cold War.[31]

Roadblock Programs

Not only did the two-war strategy generate and perpetuate some platforms that may be ill suited to the future security environment, but these systems, by eating up large chunks of scarce funds, are posing a barrier to carrying out the transformation activities that are required. The largest "roadblock" systems include, among other platforms, the tri-service Joint Strike Fighter, the navy's new carriers, and the army's self-propelled *Crusader* Howitzer.[32] Funds spent on the F-22 have already forced the air force to cancel some future modernization programs.[33] The National Defense Panel report of 1997 stated plainly that the two-war strategy was fast becoming an inhibitor to reaching the capabilities that the United States will need in the 2010 to 2020 time frame.

Increased Operational Tempo

Increased operational tempo has also created budgetary pressures against force transformation and constitutes an important reason the Pentagon has been slow to exploit the RMA. During the forty years of the Cold War, the United States deployed its troops overseas ten times, including during the Korean and Vietnamese Wars. By contrast, in the ten years following 1989 it did so on thirty-six occasions. Thus, while RMA proponents have argued that the post–Cold War era is one of "strategic pause" that can best be used to advance the RMA, to a U.S. military experiencing a very high tempo of operations "the notion of strategic pause [is] unconvincing."[34] The dramatic rise in the rate at which forces have been deployed has forced the Pentagon to devote a growing share of its funds to current activities and readiness instead of to modernization.

Higher readiness demands have compounded the problem of aging equipment, a further budgetary factor working against force transformation. Throughout the 1990s the Pentagon was faced with a reduction in defence spending and, by extension, in the money spent on procuring new systems. Even today, after the sustained budget increases that began in fiscal year 2000, procurement spending stands at only 20 percent of the Pentagon's budget, in contrast to the historical norm of 25 percent.[35] When combined with increased operational tempo, the result has been what Jacques Gansler, former under secretary of defence for acquisition and technology, has described as a "death spiral" in which the "requirement to maintain our aging equipment is costing us more each year. But we must keep this equipment in repair to maintain [current] readiness. It drains resources we should be applying to modernization of traditional systems and development [and] deployment of new systems."[36] Estimates range anywhere from an additional $25 billion to $100 billion a year that is needed just for the Pentagon to field the force outlined in the QDR of 1997, let alone to modernise and transform its forces for the future.[37]

The increased number of U.S. force deployments is a reflection of America's post–Cold War sole-superpower status. In the 1990s the United States maintained its global military obligations as the guarantor of peace in the Asia Pacific region, the Middle East and Europe, sizing its forces for a major theatre war in the Persian Gulf and on the Korean Peninsula. At the same time it adopted a somewhat interventionist policy with respect to intrastate conflict, becoming involved not only in conflicts such as those in the Balkans, which pose a potential threat to European stability, but also in missions in Africa that have a purely humanitarian dimension. Thus, the strategic realities of being the only global power is a final and important explanatory factor in seeking to understand America's slow move towards transformation.

Potential for Change

Whether or not transformation actually takes place over the next two to three decades will depend on the balance that is ultimately struck between ongoing institutional barriers to change and a growing political will for change both before and after the terrorist attacks of 11 September 2001. Institutional barriers centre primarily on the QDR process and whether or not it is capable of producing real innovation. Congressionally mandated legislation requires that the DOD conduct a defence review every four years, in the first year of a new administration. This deadline, so early in a new administration, dictates that much of the leg-work must be done before a new president is elected and well before a new administration is in place. For QDR 2001, for example, the military leadership in the Joint Staff began preparations in the fall of 1999, setting up panels to address such things as strategy, modernization, force structure, readiness, and infrastructure.

The early deadline also means that it is highly unlikely that the DOD's new civilian leadership will be able to direct the formal QDR process from the outset. While the process must begin within the first month or two of a new administration, Congressional approval of political appointees can take many months. The U.S. Commission on National Security/Twenty-first Century has argued that the QDR deadline has the effect of not allowing an incoming administration the opportunity to influence its outcome or gain a stake in its conclusions. It therefore recommends that Congress should move the QDR requirement to the second year of a presidential term.[38]

Others go further, arguing that regardless of the submission deadline, real change cannot come from the "excruciating bureaucratic process"[39] of the QDR, which must almost inevitably devolve "into a turf battle over future roles and missions – and money."[40] Nor will concrete change come from external studies by commissions, like the U.S. Commission on National Security/Twenty-first Century, that are not given "any real authority or responsibility." Rather, "crafting a new strategy requires a small civilian and military staff working directly for the secretary of defense, protected from bureaucratic pressures and willing to make some hard calls."[41]

A Presidential Boost

Such a boost seemed to be exactly what the United States was getting from U.S. president George W. Bush when he entered office. During the election campaign Bush strongly backed the idea of a revolution in military affairs, indicating that his administration would "skip a generation of technology" to transform the armed forces. By this he meant cancelling or scaling back some major weapons systems in development in favour of devoting these funds to more futuristic systems. Bush also described a future vision for the military services that is directly in line with the creation of an RMA force. "On land, our heavy forces must be lighter. Our light forces must be more lethal. All must be easier to deploy. And these forces must be organised in smaller, more agile formations. On the seas we need to pursue promising ideas like the arsenal ship. In the air, we must be able to strike from across the world with pinpoint accuracy – with long-range aircraft and perhaps with unmanned systems."[42]

Early in his term, the president tasked Secretary of Defense Donald Rumsfeld to carry out, in short order, a far-reaching strategic review of America's likely adversaries, the nature of future wars, how many conflicts the United States should be prepared to fight, and what sorts of forces it would require. Bush explicitly gave Rumsfeld a "broad mandate to challenge the status quo" and followed these words up with concrete action, pledging new spending for unmanned aerial vehicles and other "futuristic" military technologies that he expects will soon overshadow traditional weapons systems like tanks, fighters, and aircraft carriers[43]. He also promised to "transform the tradition-conscious military to deal with 21st century threats."[44] Taken together, these statements and actions gave strong evidence that the Bush administration was "moving towards what may yet turn out to be the long-promised transformation of the American defence system from a cold-war fighting force to the high-tech army of the future."[45]

But this zeal for change could not help but come up against the same bureaucratic, political and financial restrictions faced by the previous administration. The review, initially slated for completion by the end of March 2001, took longer than expected and was eventually folded into the QDR that was released on 30 September 2001. As the months wore on it became clear that Rumsfeld faced an uphill battle with Congress, civilian secretaries and uniformed officers alike to carry out even modest cuts in troops and traditional weapons systems.[46] At the same time, the administration's $1.3 trillion tax cut, combined with the economic slowdown, served to reduce the funds that were available and necessary for force transformation.

The QDR of 2001

The upshot is that the strategic review, as represented by the QDR of 2001, is less far-reaching than originally anticipated. The QDR retains current force levels, with no cuts in the number of navy aircraft carriers, army divisions, or air force fighter wings. It does not discuss specific weapons systems, either in terms of scaling back or cancelling "roadblock" programs or promoting platforms that are considered to be part of the revolution in military affairs. The overall transformational path that it outlines for the U.S. military is one that strikes a balance "between the need to meet current threats while transforming the force over time" – not unlike the course of action chosen in the QDR of 1997. As was the case with its predecessor, the QDR of 2001 could not escape the reality of needing to recapitalize some of today's legacy forces and provide for near-term readiness. Rather than skipping a generation of technology, transformation is to proceed by infusing weapons that are currently being developed with newer technologies.[47]

That said, the review does contain some important elements that are likely to promote the transformation of u.s. forces. Most significantly, it drops the force-sizing framework for two major theatres of war in favour of generating forces that can win one major war decisively while repelling aggression in another theater, in overlapping time frames. This new approach to force planning should allow the u.s. military to scale back its traditional forces and channel the savings into transformation activities.[48] In a notable departure from previous reviews, the approach also requires the United States to maintain and prepare its forces for a limited number of smaller scale contingencies in peacetime, preferably in concert with allies. This may alleviate some of the budgetary pressures related to the increased operational tempo brought on by such contingencies.

Second, and with the rise of China as a potential strategic competitor in mind, the review includes an increased focus on the East Asian theatre of operations. Experts have argued that winning a conflict in this area "will mean long-range warfare, with dispersed, mobile or concealed basing, and the kinds of forces that can sustain a long clash in the air, at sea, and in space."[49] Such requirements point away from traditional systems like aircraft carriers, fighters, and main battle tanks and towards RMA systems like the *Arsenal* ship, the stealthy B-2 bomber, and unmanned combat aerial vehicles.

Finally, and related to the new strategic focus, rather than preparing for specific threats in the Persian Gulf and on the Korean peninsula the review adopts a "capabilities-based approach" to force sizing. The capabilities that are mentioned include several RMA-related technologies and doctrines, such as advanced remote sensing and command and control, long-range precision strike, and rapidly deployable, highly lethal, and sustainable forces, including special operations forces. Strengthening joint operations also receives particular attention in the QDR as a pillar of force transformation.

Terrorism and Force Transformation

More so than the QDR, the tragic events of 11 September 2001 are likely to speed up the transformation of the u.s. military. In chapter 2 I noted that there is no common time frame for revolutionary change; rather, the rate of transition is typically a function of the level of competition in the international system. The ten years following the collapse of the Soviet Union, defined only by the vague and increasingly antiquated phrase "post-Cold War era," was a period devoid of a global organizing principle, or international paradigm, around which the United States could centre its foreign and defence policy. The terrorist attacks of September 2001 brought this "interwar" period abruptly to an end, significantly increasing the level of competition in the international system. Just as the four-decade struggle against communism provided a clear-cut means for identifying political friends and foes, so too does the new and – many anticipate – long war against terrorism.[50] Senior administration officials have already indicated that countries are either with or against the United States in this battle. And just as the Soviet threat provided a concrete focus for military force planning, so too will terrorism and the nature of the international terrorist threat.

Simply put, the United States can be expected to speed up those elements of the U.S. military's force transformation program that fit with or advance America's ability to combat terrorism. Many elements are relevant here. They include, above all, developing the smaller, more rapidly mobile, deployable, and lethal ground forces that have figured centrally in RMA doctrine from the outset. A particular emphasis is placed on special operations forces. However the force transformation efforts begun by the U.S. army in 1999 will also be essential. Not surprisingly, the QDR of 2001 calls on the secretary of the army to accelerate the introduction of forward-stationed interim brigade combat teams. In addition, the army is exploring ways it can accelerate the development of its future combat systems.[51] Strategic sea and air lift will also be important, as will combat helicopters for battlefield mobility. Heavy platforms, like main battle tanks, are likely to become even more outdated in the new strategic environment.

A second key RMA capability central to the war against terrorism is long-range precision strike. Associated platforms and weapons include stealthy B-2 bombers equipped with satellite-guided joint direct attack munitions, B-1 bombers equipped with satellite-guided air-launched cruise missiles, and submarines equipped with satellite-guided *Tomahawk* cruise missiles. Short-range tactical aircraft, dependent as they are on overseas bases, carriers, and refuelling aircraft, are less likely to be a platform of choice for military planners and political leaders.

Finally, combatting international terrorism will depend to a significant degree on advanced battlespace awareness and control capabilities. "Our highest priority right now is situational awareness," argued one high-level Pentagon official in the weeks following the terrorist attacks of September 2001.[52] Unmanned aerial vehicles like the *Predator* and the *Global Hawk* will be particularly important, as will advanced command, control communications, computing and intelligence (C4I) systems. Consistent with these trends, the Pentagon is using its share of the emergency funding provided after the terrorist attacks on New York and Washington to accelerate the development of unmanned aerial vehicles, precision munitions, and C4I programs.

CONCLUSION

Thus an assessment of U.S. efforts at force transformation gives a mixed review as to whether or not a revolution in military affairs will come to fruition over the next two or three decades. Certainly, the key technological, doctrinal, and organizational tenets of an RMA can be easily found in official U.S. defence policy and in service vision statements. Moreover, the services are taking some concrete steps, most recently organizational steps, to transform themselves. But current equipment trends indicate that America's military is not so much transforming its forces as pursuing a more technologically advanced means of "fighting the last war."

The strategic review commissioned by President Bush soon after he took office offered early promise that the U.S. military would be compelled in the not-too-distant future to address and overcome institutional barriers to change. However, the review came up against bureaucratic, political, and financial restrictions and, as expressed in the QDR of 2001, was ultimately less far-reaching than originally anticipated.

That said, the QDR does contain some signficant departures from the past that are likely to promote the transformation of U.S. forces. Moreover, the events of 11 September 2001 are likely to speed the development of several technologies, doctrines, and organizational concepts that are relevant to the RMA. The evolutionary steps towards force transformation that were undertaken in the relatively nebulous post–Cold War period have real potential, in this new era of international competition, to culminate in revolutionary change.

CHAPTER FOUR

Allied Approaches to the RMA: Britain, Australia, France, and Germany

Although today's ongoing revolution in military affairs is a U.S.-led phenomenon, many of America's allies are also taking measures to respond to the changing nature of warfare. Great Britain is, perhaps, at the forefront of these efforts. In addition, significant changes are underway in Australia, France, and, to a lesser extent, Germany. Before thinking about Canada's own approach to the challenges and opportunities presented by the "new way of war," it is useful to examine how these countries are responding to the RMA.

BRITAIN AND THE RMA

Over the past several years Britain has undertaken several initiatives that are consistent with the revolution in military affairs. A major incentive behind these efforts has been the desire to be able to provide an effective contribution to U.S.-led international security operations and thus to remain a serious player on the world stage and a valued ally. Many of the resulting decisions were captured and advanced by the *Strategic Defence Review* (SDR) released by Britain's Labour government in 1998.

Technology

A key area of focus for all the British military services is to enhance their ability to apply long-range precision force using precision-guided munitions. The Royal Air Force and Royal Navy are arming, respectively, their *Tornado* and *Harrier* aircraft with the *Storm Shadow* long-range cruise missile, which will also eventually be deployed on the *Eurofighter*. The air force is looking at acquiring a precision weapon similar to America's Joint Direct Attack Munition, which is guided by the Global Positioning System of satellites and proved highly effective during the Kosovo and Afghanistan operations. The navy is equipping its nuclear submarines, and potentially its warships, with the satellite-guided *Tomahawk* cruise missile. And the army has launched its Indirect Fire Precision Attack program to increase the range and effectiveness of its future artillery systems.

The SDR places particular emphasis on pursuing advanced battlespace awareness capabilities, which it sees as vital not only for combat operations but also increasingly for peace support missions. From 2003 the Royal Air Force expects to deploy the Airborne Stand-Off Radar, a battlefield-surveillance radar system similar to America's Joint Surveillance Target Attack Radar System, which will give the British military a high-altitude radar platform capable of gathering intelligence imagery at standoff ranges in excess of 250 kilometres. The British army is increasing its battlespace awareness by "digitizing" the battlefield, with the first battlefield digitization systems due to enter service in 2002. It is also developing the *Watchkeeper* unmanned aerial vehicle (UAV), a "ground-breaking" surveillance and reconnaissance platform that operates on a "system-of-systems" basis by merging force-level, divisional-level and unit-level requirements.[1] Although *Watchkeeper*'s current in-service date is 2008, ministry officials would like to move this up to 2004.[2] Finally, in 2000 the Royal Navy signed on to the U.S. Navy's Co-operative Engagement Capability program as a first step towards achieving a Network Centric Warfare capability (discussed in chapter 1).

The SDR also emphasises improved battlespace control capabilities. Priority is placed on ensuring Britain's command and control arrangements can work effectively in multinational operations, including NATO's new combined joint task forces, U.N. missions, and ad hoc coalitions. To this end, Britain is further developing its Joint Command System, including the provision of a secure intranet for deployed operations. Other planned improvements involve upgrading satellite communications, enhancing the joint capability of the Royal Navy's Command Support System, and acquiring the Joint Tactical Information Distribution System/Link 16 to ensure that datalinks are interoperable with those of the United States and other allies.[3] The Ministry of Defence has also identified the need to improve secure communications technology at the strategic and operational level.

Britain intends to increase its power projection capability in part by using stealth technology, an area in which Britain and the United States have co-operated for over three decades. In recent years, Washington has expanded London's access to classified u.s. stealth technology to include the successor aircraft to the Royal Air Force's *Tornado* strike interdiction aircraft. The outcome of the talks is expected to play a key role in the direction of Britain's Future Offensive Air System program to replace the *Tornado*.

Doctrine

The SDR set in train a shift from a continental European strategy to one of expeditionary forces for power projection in peacekeeping and peace enforcement operations. In Britain's view the most likely conflicts of the future will not be confrontations between large, heavy forces but regional contingencies ranging from near-combat peacekeeping operations to humanitarian missions.[4] In an environment where "the crises no longer come to us; we must go to them," deployability will be key.[5] Accordingly, Britain's strategic airlift capabilities are being significantly increased. The SDR called for Britain to fulfil its short-term requirements by acquiring four large C17 aircraft "or their equivalent" and to fulfil its long-term requirements with a replacement for its remaining, ageing transport aircraft. Since that time, Britain has leased and taken delivery of four of America's C-17s and has signed on to purchase at least twenty-five of the multinational A400M Future Large Aircraft being built by France's Airbus Industrie. The aircraft is due to enter service around 2007. However, current design specifications indicate that it will not be able to transport "outsize" equipment (as the C-17 is able to do).

As for sea lift, the Royal Navy is building two new aircraft carriers, to be operational in 2012 and 2015, each with displacements of between thirty thousand and forty-thousand tons and the capacity to carry up to fifty aircraft – including combat/strike fighters, airborne early warning aircraft, and antisubmarine helicopters. They are to replace the navy's three twenty-thousand-ton *Invincible*-class carriers, each of which carries a maximum of only twenty-four aircraft. The new, larger aircraft carriers are meant to enhance expeditionary warfare and joint capabilities, with the idea being to give Britain a fully independent capability to deploy a powerful combat force to potential trouble spots around the world. The carriers are to be equipped with the Joint Strike Fighter.

The navy is also increasing the number of its roll-on/roll-off container vessels from two to six. These vessels, which can transport heavy equipment to any port, even where sophisticated facilities are not available, will significantly increase the Army's rapid deployment capability. Each can carry half a division of troops, as well as twenty-five main battle tanks and twenty-four light armoured vehicles, such as personnel carriers and reconnaissance vehicles.

The army's rapid deployment capability is also being improved by restructuring its units into more flexible, mobile brigades. The SDR concluded that Britain's armoured regiments were smaller than the ideal and, in particular, did not contain sufficient manpower for some roles. Whereas yesterday's ground forces were structured for a short, high-intensity war in Germany, today's forces may be deployed overseas for several months or longer. To increase sustainment capabilities, the army's eight regiments have been reorganised into six larger mobile brigades. These in turn have been divided into two balanced warfighting divisions.

Although British armoured regiments are becoming larger, they are also becoming more strategically mobile, with a reduction in the number of their main battle tanks. The SDR calls for each to hold only thirty tanks for peacetime training, substantially less than the thirty-eight that had previously been planned, and roughly half the fifty-seven-tank strength of each armoured regiment in 1990. The fleet is still to be made up of the seventy-ton *Challenger* 2 tank, which entered service only in 1995. But the army is looking to replace these tanks with a variant of the Future Rapid Effect System, a family of vehicles that could enter service as early as 2007 or 2008. The twenty-ton tank version will incorporate plastic armour and stealth technology and move twice as fast as the *Challenger* 2.

Central to Britain's plans for increasing its expeditionary capability is the creation of a Joint Rapid Reaction Force. This pool of "powerful and versatile" units from all three services became operational in 2001. It includes roughly 4 ground force brigades,

20 major warships, 22 other vessels, 110 combat aircraft, and 160 other aircraft that are available in two echelons of "very high" and "high" readiness.[6] The force also includes a highly mobile air-manoeuvrable brigade based on 67 *Apache* attack helicopters, deliveries of which have begun and are due to be completed in 2003. The number of ships, battalions, aircraft, and other units that make up any particular force package will depend on the military task at hand.

The SDR set in train the creation of several other new joint organizations to maximise operational effectiveness. They encompass a joint helicopter command, merging the combat and combat support helicopters of all three services; a joint army–air force ground-based air defence organization, bringing together existing missile units; a Joint Force 2000 of navy and air force *Harrier* jets, creating a force that can operate equally effectively from aircraft carriers and land bases; and a new post of chief of defence logistics, merging the three separate service logistics organizations.

Beyond joint and expeditionary doctrine, current doctrinal trends in United Kingdom defence policy include a long-term move from manned to unmanned aircraft for combat. The Future Offensive Air System program to replace the *Tornado* strike aircraft encompasses concept definition studies not only for a new manned aircraft but also for unmanned combat air vehicles. British defence ministry and industry officials have said in the past that they consider combat UAV technologies to be most promising for the study.[7] Moreover, the SDR states that Britain will consider replacing the *Tornado* with standoff missiles or unmanned aircraft. That said, there is some question whether unmanned combat aerial vehicle technology will be sufficiently advanced to replace the *Tornado* in the timeframe necessary.[8]

In keeping with the new security environment, the Royal Navy is moving away from large-scale maritime warfare and open-ocean operations in the North Atlantic and towards littoral warfare and force projection in support of land operations. Such tasks could range from the evacuation of citizens from an overseas crisis to major warfighting operations as part of a joint force.[9] The decision to purchase two new, large aircraft carriers for expeditionary operations itself represents a move away from the antisubmarine warfare mission that dominated the Royal Navy's thinking during the Cold

War. In addition, the Royal Navy's nuclear-powered attack submarine force has been reduced from twelve to ten, with all ten submarines slated to be equipped with the land-attack *Tomahawk* cruise missile by 2008. Some high-ranking navy officials have argued that its future frigates should also have a land-attack cruise missile capability to complement the nuclear submarines.[10]

Organization

Although the revolution in military affairs predicts smaller armed forces, this is best understood as a baseline taken from Cold War levels. Britain's "Options for Change" force structure review of 1990 resulted in more than a 30 percent cut in the strength of the regular army and the Territorial Army (army reservists) and the withdrawal of more than half of Britain's forces in Germany. Its tank fleet has been reduced by 45 percent and its submarine force by 57 percent, and its air forces are being completely withdrawn from Germany.[11] Accordingly, the SDR did not call for any further reductions in the size of Britain's regular force. Indeed, the program actually represents an increase in active personnel numbers. As a result of the SDR's restructuring, the British army's regular strength has been increased by some 3,300 personnel in such critical areas as logistics and medical services. Moreover, there will be no change in the size of the regular infantry, which Britain expects will continue to be in high demand.[12]

The SDR does signal a shift towards a more highly trained force. Although the regular force will remain at essentially constant levels, the Territorial Army is being cut by more than 25 percent, from about 56,000 to roughly 40,000 personnel. In fact, the defence staff calculated an operational role for only about 7,500 ground reserves, but the government decided to retain the higher levels in order to maintain the traditional reserve force link with civilian communities.[13] The remaining Territorial Army units are being integrated more closely into the new combat capability. While the previous force was largely made up of formed units at low readiness, future Territorial Army units are to be "more readily deployable and usable" – that is, fully manned, trained, and equipped to the standards needed to be quickly incorporated into regular force elements.[14]

AUSTRALIA AND THE RMA

Over the past several years, Australia has made a conscious effort to focus on the revolution in military affairs and its implications for the Australian Defence Force (ADF). It is motivated in this effort by the desire both to maintain its "qualitative edge in the Asia-Pacific" and to remain "a highly valued ally to the United States."[15] *Australia's Strategic Policy*, released late in 1997, placed the RMA at the top of the force structure development priorities of the Australian Defence Force and emphasised the imperative of taking measures to maximise interoperability with the United States. In 1999 Australia created an Office of the Revolution in Military Affairs at the one-star level as a locus for experimenting with advanced warfighting concepts and for developing its RMA partnership with the United States. It also established an RMA Working Group to assist in refining future warfare concepts. These initiatives were complemented by the work of the Defence Science and Technology Organization of the Australian DOD on systems and technology research related to the RMA and by bilateral U.S.-Australia agreements to enhance joint capabilities and defence technology co-operation. Australia's long-awaited defence white paper, *Defence 2000: Our Future Defence Force*, released in December 2000, highlights the necessity of continuing to exploit the RMA's opportunities "in a sensible and structured way" and contains many elements that are central to the RMA's technological, doctrinal, and organizational trends.[16]

Technology

Australia's Strategic Policy states that Australia's first force-structure development priority is to exploit "the knowledge edge," that is, the effective application of information technologies that will allow Australia to use its relatively small forces to maximum effectiveness. A key part of this, stated explicitly in *Defence 2000*, is maintaining a "capability edge" in intelligence, surveillance, and reconnaissance. To this end Australia is developing a comprehensive surveillance system that can provide continuous, real-time, all-weather detection and identification of aircraft and ships in Australia's maritime approaches. It is improving its access to space-based surveillance capabilities, acquiring four Airborne Early Warning and Control aircraft, and participating in a joint U.S.-Australian evaluation program to determine the effectiveness of the *Global Hawk* long-endurance UAV in conducting standoff reconnaissance and surveillance of littoral and maritime environments.[17] The Australian Army and the Royal Australian Navy are also looking to UAVs to fulfil various roles, with the army's new tactical UAV due to enter service around 2007.

In 1996 Australia undertook a major program to improve its battlespace control capabilities. The fifteen-year Takari Program, launched by the Defence Science and Technology Organization, comprises a group of separate but related research and development projects running to 2010. Harmonising the Australian Defence Force's various efforts in the fields of command, control, communications, and intelligence collection (C3I), its objective is "to deliver a viable and integrated C3I capability to the Australian Defence Force for operations in the battlespace of the future."[18] A key step in doing so is Australia's decision, announced in its defence white paper of 2000, to proceed with the acquisition by 2006 of four Airborne Early Warning and Control aircraft and to leave open the possibility of purchasing three more later in the decade. Additional projects include higher-capacity satellite communications based on a commercial provider, enhanced broadband communications with ships at sea, and improved battlespace communications for air and land force elements. Over the next few years Australia plans to undertake a major program of improvements in intelligence support, communications, and command systems specifically for deployed forces.[19] For enhanced, interoperable command, control, communications, computers, and intelligence processing (C4I) capabilities, Australia's frigates and fighter aircraft are to be upgraded with the Link-16 datalink.

Australia is also pursuing advanced technology in the area of precision weapons and has made developing a capacity for long-range strike operations a key area of priority.[20] In 1998 and 1999 the Australian Air Force ordered two successive batches of the American-made AGM-142 missiles for its F-111s, intended for shorter-range targets like hardened bunkers. In addition *Defence 2000* announced plans to acquire new standoff weapons, the Joint Air-Surface Stand-Off Missile, which are due for delivery around 2006 and will give the F-111s a much longer-range strike capability against maritime targets in a littoral environment.

Doctrine

In the mid-1990s Australia began a process of creating mobile army task forces for the defence of its vast northern territory. *An Australian Army for the 21st Century*, released in 1996, outlined an operational concept involving self-reliant forces with individual force elements capable of independently performing a wide variety of tasks in dispersed areas. Each task force would have fully integrated supporting and combat units and greater firepower and mobility than the army's Cold War units. Each would be equipped with advanced sensors and information technology, their mobility would be enhanced with new helicopters and armoured vehicles, and they would be designed to operate jointly with the combat elements of the other services. Progress in this area has resulted in a major restructuring that has significantly increased the army's readiness and mobility. The service has also created a deployable joint force headquarters, based on the former First Division headquarters.

Defence 2000, which was greatly influenced by Australia's operations in East Timor in 1999 and 2000, broadened the army's new "expeditionary" concept to reflect "a wider range of possible contingencies, both on Australian territory and beyond."[21] Emphasis is placed on developing a professional, well-trained, well-equipped force that can be rapidly deployed and sustained over extended periods. The forces are to be organised into three brigades, each comprising two infantry battalions and a range of specialised combat and combat-support units and with much of the personnel held at thirty days to move. The overall objective is to be able to sustain a brigade deployed on operations for extended periods and, at the same time, a battalion group ready for deployment elsewhere.

Equipment acquisitions in support of a ground force with greater battlefield mobility include two squadrons of armed reconnaissance helicopters that are planned to enter service beginning in 2004 and an additional squadron (twelve aircraft) of troop-lift helicopters by 2007. One option for the combat helicopter may be America's AH-64D *Apache* helicopter. To boost force deployability, Australia has expressed an interest in procuring Airbus Industrie's A400M Future Large Aircraft, although this is very preliminary and was not mentioned in *Defence 2000*.[22] The army's M113 armoured personnel carriers are undergoing significant upgrades and eventually are to be replaced with lighter platforms under the Australian Light Armoured Vehicle program. But Australia appears to have decided against the development of a main battle tank akin to Britain's Future Rapid Effect System or America's Future Combat System, which are likely to be central to the highly lethal yet rapidly deployable high-intensity ground forces of the future.[23]

In contrast to maritime doctrinal trends in the United States, which see a move from open ocean warfare to naval power projection onto shore, *Defence 2000* makes explicit Australia's intention to retain a blue-water capability based on surface combatants and submarines. To this end the Australian navy is fully upgrading its *Collins* class submarines. That said, it is also enhancing its capability to operate in the littorals and, by extension, is promoting a doctrine of jointness. "Ours is essentially a littoral environment," Australia's chief of the navy has argued, and "not only is Australia an island, but we also need to operate within the region."[24] Australia has increased its amphibious lift capabilities with the introduction into service of the amphibious support ships HMAS *Manoora* and *Kanimbla*, from which the army's new transport helicopters will be able to operate. The two ships have undergone extensive modifications, including the installation of state-of-the-art communications systems that provide links between land and sea force elements, to give the Australian Defence Force a deployable headquarters capability. Plans are on the books to replace these ships by 2015, by which time the navy will be capable of carrying, deploying, and supporting an entire integrated battalion group.[25]

Finally, Australia is exploring the idea of unmanned combat as a possible future force doctrine. Both its F/A-18 *Hornet* air combat fighter and F-111 long-range ground-strike bombers are undergoing upgrade programs that will keep them operational until about 2015. *Defence 2000* makes provision for acquiring up to one hundred new combat aircraft to replace both the F/A-18 and F-111 fleets beginning in 2012, of which roughly seventy-five will fill the combat role and twenty-five will be optimised for strike capability. When Australia's long-term program to replace these aircraft began in 1999, the air force did "not rule out" buying a mix of manned aircraft, uninhabited combat air vehicles, and cruise missiles to replace the long-range strike capability of its F-111s.[26] This option was reaffirmed in the white paper of 2000.

Organization

Contrary to the more common organizational trends of the RMA, which see a move towards all-professional armies, the Australian Defence Force, and especially its army, is to be manned by a mix of professional and reserve forces. *Australia's Strategic Policy* noted that full-time service is increasingly necessary to develop and maintain the specialist skills needed to operate advanced systems. But it also argued that in a more flexible labour market, where people increasingly move through different careers in their working life, reserve service can be a highly cost effective way of retaining access to skilled personnel who no longer wish to continue as part of the permanent force. *Defence 2000* reiterated that "Increasingly, the Reserves will provide those skills not held within the permanent forces or held only in small numbers. The contribution of the Reserves will be essential to the maintenance of the ADF's operational capabilities."[27] The reserve component of the ADF is to be better equipped and trained than in the past and wholly integrated with the regular force component. The proportion of full- and part-time personnel in a unit will depend on its mission and readiness, but all units will contain some part-time personnel.[28]

FRANCE AND THE RMA

The term "revolution in military affairs" has only recently gained currency in France. Throughout the 1990s the RMA did not shape public debate on defence issues, and intellectual inquiry into the phenomenon was confined to military institutions of higher learning and a few branches of the Ministry of Defence. Although reflection on the RMA was slow to start in France, it is now catching up as its implications have begun to become apparent to French decision makers.[29]

Over the past few years France has carried out several significant changes in its military system that are consistent with theories about the requirements of the RMA. According to David Yost, a French defence policy expert at the U.S. Naval Post-Graduate School in Monterey, these changes have not been inspired by some "purposeful grand design" regarding the requirements of the RMA. Rather, they have been made in response to a number of distinct events and perceived necessities.[30] They include

- a desire on the part of France to define itself as a great power, to have as much national autonomy as possible from the United States, to honour its international obligations, and to uphold its international rank and standing;
- "lessons learned" from the Gulf War, including the importance of satellite intelligence, as well as the need to improve the French military's interoperability with principal allied forces, increase its power projection capabilities and bring about a better balance between conscript and professional forces;
- the requirement to adapt to new geostrategic circumstances and especially to be able to carry out crisis intervention operations in regional contingencies; and
- economic and budgetary constraints that made the cost of maintaining the French military force posture unsustainable and subsequently led to France's *rapproachment* with NATO.

Organization

In February 1996 French president Jacques Chirac announced a long-term plan for downsizing and modernising the French military. Known as Model 2015, Chirac's plan is being implemented through a series of three Defence Program Laws, the first of which covers the period 1997 to 2002. The most dramatic aspect of Chirac's announcement and of the subsequent Program Law was the decision, after over two hundred years, to all but eliminate conscription and thus professionalise the French armed forces. By 2002, 92 percent of the French military will be professional, as compared to 59 percent in 1996.[31] Key to this process is a significant reduction in force levels. The French armed forces are being reduced from a 1996 level of 500,000 to 357,000 personnel by the end 2002.[32] This represents a 28 percent reduction in force size – a proportion comparable to the reduction in British force levels since 1990. Of this figure the total reserve manpower will remain at 100,000 personnel. However, rather than providing formed units, reservists are to be employed as individuals to augment regular force units. Accordingly, all reserve brigades and reserve regiments have been disbanded.

The French army is undergoing a significant organizational transformation. By the end of 2002 the total number of regiments will have fallen from 129 to 85, including 51 manoeuvre regiments,

15 logistics regiments, and 19 specialised support regiments. These, in turn, are being grouped under ten manoeuvre brigades (including the Franco-German brigade) and six specialist brigades.[33] Despite this grouping, the regiments themselves are meant to be the basic modular units of the army that can be brought together in varying combinations, depending on the nature of the crisis to which they are responding. This notion of "modularity" extends not only to the French army but also to the other services, the idea being to create a versatile force capable of joint tailoring and combined operations.

France's decision to professionalise its armed forces was based on the demands of future warfare and the nature of the post–Cold War security environment. The experience of the Gulf War demonstrated that the short-term military service of a conscript was no longer compatible with the high degree of technical proficiency, tactical and operational training, and extensive experience required of military forces in the coming decades. Moreover, it brought home the fact that France had very few high-quality professional forces to contribute to a coalition operation far from the Cold War's central front. Both during the Gulf operation and later with respect to the rapid reaction force that went into Bosnia in 1995, France could deploy only about 10,000 troops by "skimming off" professional soldiers for such duties. Chirac recognised that the most likely challenges of the future would be interventionary operations far from home and that it would be necessary for France to have professional soldiers prepared to conduct such operations.

Doctrine

The Program Law of 1996 stressed the importance of the French services being able to operate together and of France being able to take part in combined operations with its allies. Thus the French services have been reorganised with greater modularity, so that their units can be more easily formed into a joint force and combined with elements of Britain's Joint Rapid Reaction Force and Germany's Crisis Reaction Force (discussed below). In addition, four joint force headquarters are being created in place of the division-level commands, each of which will be able to take command of a 5,000-strong French force or of a NATO force of between 12,000 and 18,000 personnel. These four staffs do not have permanently assigned units under their control; rather, they will take command of whatever force is constructed for the particular task at hand. The army has also formed a new corps-level headquarters, Land Force Command, which can function as a command and control organization for a NATO Combined Joint Task Force (see chapter 5) or a multinational force equivalent to the size of a NATO corps.[34]

These moves come on top of several organizational changes France made soon after the Gulf War to facilitate joint and combined operations. In 1992 it established a Special Operations Command to bring together the special operations forces of the army, navy, and air force. In 1993 the government established a Joint Operational Planning Staff to prepare for engagements in Europe and beyond, with an emphasis on joint and coalition operations. And that same year, France established a new Joint Defense College, which replaced the war colleges of the army, navy, and air force.[35]

The Program Law notes that the kinds of conflicts France is likely to be confronted with in the future will require power projection for brief engagements using forces that are smaller in number but immediately available and operational. To this end, France is creating a Rapid Action Force of between 50,000 and 60,000 troops that will be quickly deployable to areas around the world. The objective is to be able to rapidly deploy either a force of 50,000 for NATO non-Article V contingencies or a force of 30,000 to high-intensity conflicts for up to a year, while at the same time maintaining a force of 5,000 for low-intensity combat or peacekeeping operations.

Creating this force involves not only the professionalisation, reorganization, and joint efforts noted above but also furnishing the units with new, lighter, more advanced, yet still highly lethal, equipment. Acquisitions in support of Model 2015 are projected to take place up to that date and will figure significantly in the plan's final two Program Laws. For increased battlefield mobility France is developing, in cooperation with Germany, the *Tigre* combat helicopter, with its 215 aircraft deliveries scheduled to be spread out between 2003 and 2010. The French army is also to receive about 130 NH90 tactical troop transport helicopters, capable of lifting up to twenty commando troops each or a light combat vehicle. For increased force deployability, the army is purchasing up to 700 of the

Véhicule Blindé de Combat d'Infanterie, a wheeled platform that is lighter than its predecessor (the AMX-10P) and will be used to complement the *Leclerc* main battle tank. Deliveries are planned to begin in 2005. Although the number of France's heavier main battle tanks has diminished by about 60 percent, the Program Law went ahead with the purchase of some 400 new *Leclerc* tanks, rather than focusing on the development of a lighter combat system akin to America's Future Combat System.

For increased airlift the French air force is participating in the multinational Future Large Aircraft development program and is to retain its current lift capacity at about 50 strategic transport aircraft. France is also considering leasing some C-17 aircraft to cover its short-term strategic lift gap before the A400M is fielded in about 2007. In 1999 France backed a German initiative to form a European Military Airlift Command to pool airlift resources, such as strategic transport aircraft and air-to-air refuelling tankers, across national boundaries. This initiative is now being considered in the European Air Group.

Meanwhile, the French navy is building a power projection force around the nuclear-powered aircraft carrier *Charles de Gaulle*. Commissioned in 2000, the carrier displaces 40,000 tons and carries up to 40 aircraft, including the new *Rafale* fighter and the U.S.-built E-2C *Hawkeye* early-warning aircraft. In contrast to Britain, which is building two new aircraft carriers by 2015, France has scaled back its carriers. With the retirement of the *Foch* it now has only one carrier, and it is increasingly unlikely that a sister ship, originally scheduled to enter service in 2012 to 2015, will actually be built.[36] Future navy acquisitions are to take into account that most naval operations in the 1990s took place near coasts rather than in the open ocean. According to one French analyst, this means that there is a drop in the need for destroyers and a greater need for frigates and corvettes.[37] Accordingly, the navy's power projection force is to include, in addition to the carrier, new amphibious warfare ships, new anti-aircraft frigates, and countermining ships.

Technology

Although the most dramatic changes in the French military over the past few years involve organizational and doctrinal developments,

France is also pursuing a number of technological advances that are associated with the RMA. The Persian Gulf conflict underscored most emphatically for France the value of space-based surveillance systems, especially for intelligence gathering. Moreover, it highlighted the fact that France (like other allies) was largely dependent on U.S. systems for obtaining such information. Motivated by a political determination to be able to make independent intelligence assessments, French officials argued in the months following the Gulf War that France must develop an "autonomous capacity for space observation."[38] To this end, France launched its first *Helios I* observation satellite in 1995, followed by a second in 1999. The *Helios II* infrared satellite is expected to enter service in 2004, and project definition studies have begun for the *Syracuse III* military satellite. In 2000 France and Germany began a cooperative venture to build a military radar satellite system of between four and six satellites.

More recently, France is reassessing the utility of satellite surveillance versus airborne systems. "The fact remains that satellites' role in battle is not crucial," argued French defence minister Alain Richard following the Kosovo conflict. "They are too high up. It is better to use drones or reconnaissance aircraft."[39] This reassessment will build on measures France has taken over the last several years to increase its battlespace awareness capabilities with UAVs. In 1995 it replaced its *MART* system, a mini-UAV it used during the Gulf War, with the more capable *Crecerelles*. France deployed the *Crecerelles*, as well as the Israeli-produced *Hunter* UAV, during NATO's Kosovo operation of 1999. It plans to retain the *Crecerelles* in service until at least 2005, pending entry into service of the *Brevel* twenty-four-hour unmanned reconnaissance aircraft, which it is developing with Germany.

Meanwhile, the French army is enhancing the situational awareness of its troops by carrying out a process similar to that of the United States of "digitizing" its force. By 2005 each French soldier is to be equipped with the ability to detect targets by day and night to a range of 300 metres, identify friendly troops, use remote sighting to aim and fire from behind cover, and transmit video data that is received on a helmet-mounted display. In 2004 France will contribute a complete brigade to an American-led digitization exercise. The overall objective is to create a "general state of digitization" of the army by 2010.[40]

Modernising its C4I capabilities is a top priority of the French military. The army has fielded an enhanced information system that allows for the transfer of data to multiple tactical levels in real time.[41] And to keep up with its C4I goal of continued interoperability, the navy's *Charles de Gaulle* carrier, E-2C *Hawkeye* aircraft, and *Rafale* fighters are to have the Link-16 datalink.

For precision strike, France has shown significant interest in stand-off systems and has placed them "among its highest equipment priorities."[42] The air force has begun taking delivery of the *Apache* air-launched cruise missile and is fitting it to their *Mirage 2000* fighter aircraft. In 2002 France will begin taking delivery of about five hundred *Scalp* cruise missiles, which have a range of over 250 kilometres and use the Global Positioning System for mid-course guidance updates. France is also building a new air-to-surface missile that will enter service with its *Mirage 2000D* fighters in 2004. The idea of unmanned combat aerial vehicles delivering precision munitions has also "caught on in France," with two French companies looking at UCAV options for the future.[43] Finally, France is focusing on the value of stealth across a range of platforms. Stealth technologies are being built into plans for its future frigates,[44] and France, Germany, and Sweden have undertaken a joint technology development effort to design a stealthy combat air vehicle.[45]

GERMANY AND THE RMA

Germany has only recently begun to focus on the revolution in military affairs and its implications for the German armed forces. Throughout much of the 1990s the RMA was a moot issue in Germany. It was only in 1994 that Germany overcame its constitutional constraints to deploying forces beyond its borders, and it was not until 1995 that it actually undertook such a deployment (into Bosnia). Germany's contribution to NATO's operation in and around Kosovo in 1999 represented the first time the nation had conducted combat operations since World War II.

In May 2000 the independent Commission on the Common Security and the Future of the *Bundeswehr*, which had been appointed by the federal government a year earlier, submitted its report to parliament. The report represented the most far-reaching study of Germany's national security posture in close to thirty years, and its findings included many elements that are central to the RMA. As in France, the commission's most dramatic conclusions came in the area of organization. Indeed, the commission's principle task was to draw up proposals for the basic structures of a new *Bundeswehr*.[46]

Organization

The German military undertook dramatic force reductions in the decade following unification. From a combined East and West German military personnel strength of 583,000 in 1990, the *Bundeswehr* was reduced (by the terms of unification) to 370,000 troops by 1994. Over the next five years, these levels fell still further to a peacetime strength of about 320,000. In the wake of the commission's report, Germany's military force levels are to be reduced to roughly 275,000 by about 2004 – a figure that is still somewhat higher than the commission's recommended level of 240,000.

In contrast to the more common organizational trends of the RMA, Germany continues and likely will continue for the foreseeable future to have a system of conscription. During the first post–Cold War decade, successive German governments stayed clear of any talk of reducing or eliminating conscription, arguing that national service was (and is) an important part of German defence culture that effectively binds the military and civilian society together. In the wake of glaring evidence provided by *Operation Allied Force* that a conscript force was ill-suited to today's international security environment, the commission recommended significant changes to the conscript system. It called for an annual figure of 30,000 conscripts on a ten-month term of service, in place of the previous level of 135,000. But it fell short of taking the French route of eliminating conscription altogether, arguing instead that "in view of enduring external uncertainties ... the *Bundeswehr* of the future cannot rely solely on volunteers." In the months following the commission's report, the German government announced that the number of conscripts would be reduced to about 80,000 per year.

Doctrine

Perhaps no Western military establishment was less suited to the rapid deployment demands of the post–Cold War era than that of

Germany. Four decades of preparing for a massive, predictable Soviet assault on the eastern border of West Germany meant that Germany had a military force with neither a rapid, long-distance deployment capability nor the logistical tail to support one. The Gulf War conflict drove home the realisation at the Ministry of Defence that Germany was completely unable to deploy its forces outside its own territory. Germany's *Defence White Paper, 1994* stressed that while the *Bundeswehr* had sufficient advanced equipment for the task of national defence, "what is still missing is an ability to participate effectively in international crisis management activities. There is a requirement for a German reaction force contingent capable of interacting with allies and of covering the entire spectrum of possible missions."[47]

In response, in the latter half of the 1990s Germany created a Crisis Reaction Force of about 50,000 troops, organised from units of all three services. The reaction force is kept at full strength and consists entirely of professional forces. It is comprised of six fully equipped army brigades, eighteen air force squadrons and roughly 40 percent of the navy's ships at any one time.[48]

Despite this initiative, defence experts continued to argue that more needed to be done. A decade after the fall of the Berlin Wall, Germany's ground forces remained largely configured to fight a defensive war on German soil, with almost 90 percent still tailored for duties as main defence forces. While the Crisis Reaction Force constituted 70 percent of the air force and nearly 100 percent of the navy, it included just 16 percent of the army.[49] "We generally lack the 'lighter' reaction forces that might be needed at the outer perimeter of NATO," argued the director of the commission in 1999. "We have about 50,000 reaction forces today; we need about three times that number."[50] Increasing the *Bundeswehr*'s rapid reaction capability figured strongly in the commission's final report and the Crisis Reaction Force is now in the process of being transformed into a 150,000-strong readiness force, with sailors, soldiers, and airmen available in three 50,000-man rotations.

The Crisis Reaction Force's deployability is to be enhanced with new, lighter combat vehicles and increased airlift assets. The German army is replacing its armoured infantry fighting vehicle with the lighter, wheeled All-Protected Carrier Vehicle, with delivery expected to begin in 2002. It has also undertaken to replace its main battle tanks with a lighter New Armoured Platform that will not sacrifice firepower or troop survivability. The army's battle-field mobility will be enhanced with the NH90 tactical troop transport helicopter. For increased strategic lift, the German air force will eventually receive 73 of the A400M heavy lift transporters. Like France, it is also considering leasing some C-17 aircraft until the Future Large Aircraft enters into service. The German navy is focusing not so much on strategic lift but rather on building a flexible naval force for "international crisis management."[51] To this end new frigates, U-2 boats, and supply ships have been approved by parliament.

Technology

In terms of RMA-related technology, Germany has focused to a significant degree on advanced intelligence, surveillance, and reconnaissance systems, particularly with respect to UAVs. Indeed, "Germany holds the title [in Europe] for examining the widest range of UAV options."[52] Since 1992 Germany has been working with France in developing the *Brevel* twenty-four-hour unmanned reconnaissance aircraft and has already begun taking delivery of these platforms. It also has seven batteries of CL-289 drones, one of which it contributed to NATO's air war against Yugoslavia. With its LUNA X-2000 Germany leads the way in developing short-range drones for brigade-size formations.[53] Germany is also participating in a NATO program to develop a Maritime Unmanned Aerial Vehicle that will provide support for surface warfare, electronic warfare, antiship missile defence, amphibious operations, and antisubmarine warfare.[54] Beyond advanced surveillance technology, Germany plans to acquire the *Taurus* standoff, precision-guided cruise missile and is taking part in a joint (with France and Sweden) technology development effort to design a stealthy combat air vehicle.

CONCLUSION

This short survey of measures that Britain, Australia, France, and Germany are taking to respond to the revolution in military affairs reveals a number of common themes. Technologically, there is strong focus on advanced intelligence, surveillance, and reconnaissance capabilities, particularly using UAVs, as well as on advanced command and control and enhanced precision-strike capabilities. Organizationally, a key trend is the move towards professional, highly trained personnel and the creation of smaller, more modular units that facilitate both joint and combined operations. An important overarching doctrinal theme – one that is influencing many technological and organizational changes, as well as some doctrinal developments – is the creation of lighter, more rapidly mobile, yet still highly lethal, forces. Expeditionary doctrine is in large measure driven by the view that most conflicts in the foreseeable future will be regional contingencies ranging from humanitarian missions to near-combat peace-enforcement operations and now to operations to combat the scourge of international terrorism. Britain, Australia, France, and Germany are therefore placing their emphasis on furthering those aspects of the RMA that are best suited not just to high-intensity conventional warfare but to a broad range of missions. That said, many of their measures are as yet aspirations – a stated policy for future direction. The next chapter takes a "snapshot in time" to examine the degree to which European members of NATO are currently capable of responding to the demands of the new international security environment.

CHAPTER FIVE

NATO and the RMA

One of the most significant outcomes of NATO's summit of April 1999 was the decision taken by alliance heads of state and governments to launch a Defence Capabilities Initiative (DCI). The objective of this initiative is to ensure that the alliance can effectively carry out operations across the entire spectrum of its present and future possible missions – from responding to a humanitarian disaster to carrying out a peacekeeping or peace-enforcement operation to conducting high-intensity warfare in defence of alliance territory. Although originally conceived in the fall of 1998 as a measure to address the growing technology gap between the United States and its NATO allies, the initiative was broadened in the months before the summit to include key doctrinal and organizational requirements of future military operations. In doing so, it went well beyond long-standing initiatives like the NATO Standardization Program to respond to central elements of the revolution in military affairs. Its timeliness and importance was soon demonstrated by NATO's Operation Allied Force in and around Kosovo in the spring of 1999.

THE TECHNOLOGY AND CAPABILITY GAP

Technology

Operation Allied Force made manifest a significant technology and capability gap between the United States and its European allies (Canada will be discussed in chapter 8). In the first instance, it revealed the limited precision force capabilities of many European members of NATO. While the alliance as a whole deployed substantial precision firepower, this was overwhelmingly due to U.S. assets. Although thirteen countries took part in the NATO air campaign, more than 70 percent of the firepower deployed was American.[1] Less than half the European members of NATO had laser-guided bombs, and only Britain was able to contribute cruise missiles.[2] Only the United States had all-weather (i.e., satellite-guided) air-launched precision munitions. In terms of aircraft, of the five thousand or so that European militaries could theoretically use for air strikes, barely 10 percent were capable of precision bombing.[3] France was the only European ally able to make a significant contribution to high-level bombing raids at night, accounting for about 14 percent of all allied strikes at ground targets.[4] At the same time, only the United States could contribute strategic bombers and stealth aircraft for enhanced power projection.

European allies also critically lacked intelligence, surveillance, and reconnaissance assets. Although the alliance has for many years operated Airborne Warning and Control System aircraft for aerial surveillance, it has yet to acquire an air-to-ground surveillance capability akin to what is provided by America's Joint Surveillance Target Attack Radar System.[5] The alliance was therefore entirely dependent on the United States for this means of surveillance. European reconnaissance and surveillance capabilities were somewhat stronger in the area of unmanned aerial vehicles. In addition to America's *Predator* UAV, the allies deployed French and German UAVs in the Kosovo operation.

Although European members of NATO are taking selective measures to respond to the RMA, they are not incorporating advanced technologies into their military systems quickly enough to stem a very evident technology gap between the U.S. military and its European counterparts. A report in 1998 by the science and technology committee of the North Atlantic Assembly stated that "there is no doubt that a transatlantic technology gap exists. Without remedial actions by the United States' allies, sooner or later – and probably sooner – the gap will become a rift."[6] Since then, the pace of the modernization of U.S. information systems has been such that the gap has continued to widen.[7]

Doctrine and Organization

Soon after the end of the Cold War, NATO began to adapt its multinational, joint military command structure to the new security environment. The old standing command structure that was aimed primarily at defending members against a massive Soviet threat was significantly modified and the transition to a new command structure begun. In the early 1990s NATO also established Allied Command Europe Rapid Reaction Corps (ARRC), a mobile headquarters of around a thousand multinational military personnel. And since 1994 the alliance has been working to implement its Combined Joint Task Force (CJTF) concept. This entails the development of mobile command and control headquarters that can be detached from the alliance's permanent standing command structure and that will allow units from different services and nations to be brought together and tailored to a specific contingency.

Although the CJTF concept is proceeding, it has yet to be fully implemented. Moreover, there is a growing recognition that the ARRC – which was deployed to both Bosnia in 1995 and Kosovo in 1999 – needs to be augmented by at least a second such mobile corps headquarters. The chairman of NATO's Military Committee has argued that the figure may be closer to three land corps headquarters and forces at high readiness and six such headquarters at lower readiness for sustainment operations. Comparable forces and headquarters are needed for the navy and air elements.

At the national level, as discussed in chapter 4, all NATO's key European members have adopted, in one form or another, force mobility and power projection as the guiding doctrine for transforming their armed forces. But despite these efforts, European members of NATO remain hampered in their ability to contribute forces suited to post–Cold War missions. Although their forces comprised almost two million people, compared to America's 1.45 million, they could draw up only half the number of properly equipped and trained professional soldiers specified for the Kosovo operation.[8] A key reason for this is the slow move towards military professionalisation and force restructuring on the part of some countries. "Europe today has more than enough forces. The problem is that European forces – Britain and France excepted – are not structured to deal with the type of security threats the Alliance is likely to face in the future."[9]

The expeditionary capability of European nations is also hindered by the fact that almost all European members of NATO have no equivalent to the C-5 and C-17 heavy transport aircraft and are therefore dependent on the United States for air transport of troops. Britain has leased four American C-17 air transporters, and France and Germany are considering doing the same. In the longer term, eight European nations have signed on to the multinational Future Large Aircraft. But delivery will not begin before 2006. Moreover, even once delivered these aircraft are unlikely to have the lift capacity to accommodate "outsize" military equipment – an important point considering the DCI's High Level Steering Group has emphasised the requirement for "strategic lift – especially outsized air transport."[10]

EXPLAINING THE GAP

For some years it has been apparent that there is a significant capability gap between the United States and its European allies. During the 1991 Gulf War, coalition partners were shown to be deficient in a wide range of areas, including intelligence, surveillance and reconnaissance, communications, precision attack, long-range transport, and force protection. Since that time the gap has only widened. The NATO peacekeeping operation in Bosnia emphasised once again the technology disparity between U.S. and allied forces in communications, intelligence, and surveillance capabilities. The Kosovo operation highlighted significant European shortfalls in precision-guided munitions and missiles, long-range bombers, and stealth aircraft.

The widening gap is partly due to substantially reduced defence budgets in all West European countries in the post–Cold War period. In the decade following the collapse of the Berlin Wall, NATO's European members cut defence spending by about 25 percent in real terms. The European allies spend on average 2.1 percent of gross national product on defence, compared to 3.0 percent by the United States, and military spending in nine European NATO countries falls below the NATO minimum recommended level of 2 percent.[11] Budgetary constraints have in turn substantially reduced the ability of European members of NATO to purchase attack and transport helicopters, air and sea lift assets, and sufficient stocks of advanced precision munitions – all requirements to address crises in the new international security environment.

It is not only reduced defence budgets that have widened the capability gap. After all, U.S. military funding was also cut by 25 percent in the 1990s. Moreover, EU defence spending at the end of the 1990s was about 58 percent of American spending, somewhat lower than the 60 percent figure of the middle of the decade but still well within Cold War norms. In the mid-1980s, for example, at the height of the "second Cold War," EU members were spending the equivalent of 55 percent of the U.S. defence budget on their militaries.[12]

Rather, the problem also lies in the manner in which remaining defence funds are put to use. Too many European resources continue to go towards expensive facilities, unnecessarily expensive weapons procurement, and conscripts that are ill-suited to today's security environment. In this sense, Europe's difficulty lies more in the way its armed forces are structured and equipped than in its overall level of defence spending.[13] Many European members of NATO do not devote enough of their defence budgets to procurement. Levels range from a planned 40 percent in France, to 15 percent in Germany, to 10 percent in Belgium.[14] All told, European allies' defence procurement accounts for less than a third of NATO's total.[15] The upshot is that while "European countries spend about two-thirds of what the United States spends on defense [they] don't have anything like two-thirds of the capabilities."[16]

In addition, as in the United States, the procurement budgets of the major allies are tied into "roadblock programs." For example, they are all slated to acquire large new fleets of combat aircraft, like the *Eurofighter*, *Rafale*, and Joint Strike Fighter, some of which had their origins in the Cold War and all of which may not be well suited to the nature of the future security environment. But unlike in the United States – where the debate often seems "only to pit those who want to spend more on defence against those who want to spend a lot more"[17] – in Europe there is not the same kind of defence budget leeway to be able to undertake such programs and still be able to devote substantial funds to a future force.

The United States has also more aggressively pursued revolutionary innovations in software, communications, sensor, and logistics technology to compensate for manpower and equipment reductions.

In absolute terms, the United States spends roughly two and a half times as much on research and development as all European members of NATO combined.[18] As a result, the allies invest far less than the United States in advanced military information systems, in research and development of new technologies in general, and in recruiting, training, and retaining high-quality personnel.[19]

A further reason behind the gap is that in the 1990s the United States undertook a wholesale restructuring of its defence industries. Huge mergers brought together a range of specialities – engineering, electronics, software design, and so on – into a few large firms, leading to both efficiencies and innovation. By contrast, European firms have not undertaken significant cross-border mergers and still restrict most of their sales to their own military establishments. The result is redundant company overheads and a lack of competition, resources, and innovation. Economies of scale also come into the picture. Many NATO countries simply lack the economic viability or large-sized military force that is required to efficiently develop highly expensive specialised capabilities, such as low-observable technologies for varying platforms.

Finally, some analysts argue that Europe's real weakness in security matters lies not in the state of its advanced military technologies but in a deep reluctance, after years of following the U.S. lead, to "think strategically."[20] A more balanced take on this basic point is that the nature of the Cold War encouraged Europeans to concentrate on "homeland defence," while geography dictated an American focus on expeditionary warfare. The end of the Cold War has therefore forced Europe to make a much bigger reassessment of its military needs.[21]

IMPLICATIONS OF THE GAP

The most immediate implication of the technology and capability gap is that European armies may soon be unable to operate alongside the Americans because of their "technological backwardness."[22] NATO secretary general George Robertson has spoken of a "two-speed" alliance, with Europe unable to keep up with the United States.[23] Similarly, high-level U.S. officials have argued that differences in technological capabilities could create an alliance with three tiers of membership: the United States, technologically capable of "doing everything"; close European allies, capable of performing many, but not all, advanced warfare tasks; and new members largely incapable of "keeping up with the pack."[24]

Already there are indications that interoperability problems are hampering the allies' ability to carry out combined operations. During the Gulf War, America's superiority in information systems and electronics meant that its military communications were much faster and more sophisticated than those of its allies. More recently, the Kosovo operation highlighted problems in joint deployment, target identification, and weapons compatibility,[25] and indicated that European air forces will likely be unable to operate alongside U.S. fighters in future conflicts unless they develop and procure new military electronics.[26] Incompatible communications links also forced American aircraft, which are equipped with secure links, to operate on the open air waves with their allies.

Although problems of compatibility have been an issue for the alliance since its inception, the difference today is that U.S. advances in communications, data processing, and precision-guided weapons are in the process of "completely eclipsing" those of its allies and casting into question their ability to function together.[27] The technology gap is making it harder for Americans and Europeans to fight together at the very time when the changing nature of warfare suggests that they should be integrating their forces ever more closely.

The capability gap between the United States and its European allies has implications beyond the creation of compatible forces. The inability of allies to work together in operations could breed new tension within the alliance, undermining not only its effectiveness but also its cohesiveness. Tension could arise if by default European armies find themselves increasingly responsible for the dangerous, manpower-intensive tasks that can lead to significant casualties. Given current technological trends, there is a risk that in future combined operations Europeans will be expected to provide the bulk of the ground forces, while the United States provides the high-tech logistics, lift, intelligence, and air power. Already we saw this sort of situation in Bosnia in the mid-1990s, when European and Canadian ground troops were deployed as part of the UN mission while the United States restricted its contribution to air strikes – which unwittingly put the soldiers at risk.

A growing technological "rift" also increases Europe's security dependence on the United States, raising burden sharing issues. "Post–Cold War U.S. defense policy calls for Europeans to do more for their own security ... But the disparities in capabilities threaten to have just the opposite effect, with the Europeans becoming even more dependent on superior U.S. military power."[28] While the divergence in capabilities has not prevented successful operations in Bosnia or Kosovo, many experts argue that European militaries could provide little help in a more demanding engagement.[29] Already we have seen evidence of this in Afghanistan. Thus, ironically, the more severe the threat to interests shared by the United States and Europe, the less likely it is that a true U.S.-European coalition will respond.

If this is true, then the technology and capability gap among allies could not only breed new tension within the alliance but, far more significantly, it could marginalise the political and military importance of the North Atlantic alliance. During the Cold War the alliance was held together by a common, massive threat. With this threat gone the new "glue" for alliance cohesion must be the ability to work together to respond to the new risks and threats faced by alliance members. Continued political and military support for the alliance on the part of the United States will ultimately depend on the ability of its European allies to make a valid contribution. "The common argument that a stronger, more assertive Europe will undermine NATO as well as U.S. interests is simply wrong. A Europe that remains allied to the U.S. because of its own weakness is of limited value in the current strategic environment, and probably unsustainable politically."[30] While NATO would not disappear overnight if its European members refused to adapt, the alliance could quite quickly lose its place in the hierarchy of the United States' overseas security commitments.

Increasing the military capability of NATO's European members is also in Europe's interest. The more substantial a contribution Europe can make to an operation, the greater the influence it will hold over American decision making in a crisis or conflict. Equal influence depends on comparable capability – and in this sense there can be no real sharing of decision making power as long as the inequality in the means to act continues to increase. Closing the technology and capability gap is thus the first step in bolstering the European pillar of the alliance, enhancing the U.S.-Europe partnership, and strengthening the alliance as a whole. Far from replacing NATO, enhanced European military forces are central to the future viability of the alliance.

The crisis in Kosovo brought home vividly the deficiencies in European military capabilities. Ten years after the end of the Cold War and three years after the end of the war in Bosnia, it was clear that Europe still relied on U.S. leadership to respond to a crisis that most directly affected European interests. Britain and France had already issued, in December 1998, a Joint Declaration on European Defense from St Malo, France, in which they agreed to develop a joint defence capability within the institutional framework of the European Union (EU) – something to which Britain had not previously been willing to agree. Alarmed by their continuing dependence on the United States, as revealed by the Kosovo crisis, EU leaders vowed their commitment to the development of an effective common European security and defence policy. To lead this drive, they appointed former NATO secretary general Javier Solana as their high representative for a common foreign and security policy. They also merged the long-dormant Western European Union with the EU and set up a number of new political and military decision-making structures. At their summit in Helsinki in December 1999, European Union leaders decided to create, by 2003, a rapid reaction force of some fifty to sixty thousand troops to carry out the "Petersberg tasks" of humanitarian and rescue missions, peacekeeping, peacemaking, and crisis management. The following November, at a military capabilities commitment conference, EU members formally pledged about a hundred thousand troops, four hundred combat aircraft, and one hundred ships to a "catalogue" of forces available for future crises.

But a viable European security and defence policy – and, by extension, a stronger European security and defence identity within NATO – will depend on more than strengthening decision-making structures and stating future commitments. Institutional changes and capability commitments are less important than actually modernising forces to meet the demands of the new international security environment. In part this means EU members spending more on defence. It also means most members spending their defence budgets on different things.

THE DCI ROLE

It is here that the NATO Defence Capabilities Initiative has a key role to play. The DCI seeks to increase NATO military capabilities in five specific areas: deployability and mobility of allied forces, their sustainability and logistics, force survivability, effective engagement capability, and their command and control and information systems. Overarching all of this, the initiative places special emphasis on improving interoperability among the forces of member states. The DCI is therefore specifically designed to address those areas where the alliance needs to develop its military capabilities in order that it can most effectively respond to the sorts of challenges it is likely to face in the coming years and decades. In this sense, the initiative is central to determining where European militaries should focus their modernization efforts.

This is true despite a longstanding debate within the alliance over the future role of NATO. At the root of the differences lies the American conviction that NATO should be seen as an "alliance of interests," as much as one dedicated to the defence of a specific territory. These interests could see NATO conducting missions in places like the Persian Gulf, where weapons of mass destruction hold the potential to threaten alliance members. European allies have questioned the U.S. position; they are fearful of watering down NATO's core mission of collective defence and wary of becoming some sort of a "junior partner" to American strategic interests. The Washington summit communiqué found a middle ground, stating that while collective defence of alliance territory remained NATO's core mission, the alliance "must also take account of the global context."[31] With some foresight, NATO's Strategic Concept of 1999 also stated that acts of terrorism could threaten common interests. Following the terrorist attacks on the United States the alliance invoked article 5 for the first time in its fifty-two-year history, clearly indicating that when it comes to the collective defence of an ally, European members of NATO are willing to undertake military operations in places that would previously have been seen to be well out of area.

Even before the events of 11 September 2001 the debate over NATO's operating area was an essentially moot point when it came to military capabilities and requirements. "The distinction between Article V and non-Article V capabilities – that is, between collective defence and crisis response – has become artificial. Modern collective defence and out-of-area crisis response share the same vital requirement: projecting combat forces rapidly over great distances."[32] Whether it be to the Balkans or the Persian Gulf, European allies will need to be able to rapidly project power, sustain forces for long periods of time, engage an adversary effectively across a wide spectrum of possible circumstances, survive the potential use of weapons of mass destruction, and have effective command and control arrangements.

Not surprisingly, these same requirements are the primary areas of focus for the DCI. Moreover, they echo key elements of the RMA. Deployability is best enhanced by investing in air and sea lift and by reorganising forces into professional units that are smaller and more modular and equipped with lighter, yet highly lethal, weapons. Sustainability will be dependent in large part on applying advanced technologies to logistic efforts. Effective engagement requires a wide variety of advanced weapons systems associated with the RMA, from precision-guided munitions and all-weather surveillance and reconnaissance systems to attack helicopters and stealth aircraft. It also necessitates that these systems be interoperable among services and militaries to facilitate joint and combined operations. For effective engagement and survivability, NATO forces need to be able to address nuclear, biological, or chemical weapons threats in theatre. Finally, advanced, interoperable, and deployable command and control and information systems are key elements of the RMA and essential for enhancing military capability.

PROSPECTS FOR IMPLEMENTATION OF THE DCI

Progress in implementing the DCI in its first few years can at best be described as fair. Formally launched in April 1999, the DCI saw little, if any, concrete progress in its first eighteen months.[33] A NATO assessment in mid-2001 reported that less than a third of the DCI's specific undertakings were in a stage of being "nearly or fully completed," and even here many of the boxes checked were linked to reaching interim achievements, with real capability improvements to come some time later. Just over a third were said to be showing "significant progress," while in the remaining areas substantive activities had not yet begun.[34] Projections are that alliance leaders will have very little in the way of success to report when they meet for their November 2002 summit in Prague.[35] In short, despite the DCI's goals, NATO members have not seen the capability gap shrink in any appreciable manner.

Whether or not the DCI will succeed in the long run is as yet an open question. Certainly, much can be done within existing defence budgets to increase capabilities. Two-thirds of the DCI's fifty-eight action items are related to forces and capabilities and can therefore be addressed through the NATO force proposal process. The challenge here is for individual nations to reprioritise defence acquisitions and decisions to accord with DCI decisions. A number of allies have already reviewed their defence plans, with a view to the more efficient use of defence resources. Indeed, it is primarily in areas related to policy, procedures, or other planning efforts that the DCI has registered progress.

Reprioritising within defence budgets can only do so much, however. In most cases the chief impediment to implementation is the fact that the specific DCI objective depends on higher defence spending and especially on the portion of a nation's spending that is devoted to equipment acquisitions. The 1990s trend towards dramatically reduced defence budgets has only recently reversed. Britain and especially the United States have registered real increases in their defence spending in the past few years. The Canadian defence budget has increased slightly, while the French defence budget has remained essentially static. Twelve other European members of NATO (including Germany) have indicated that they plan real increases in defence spending, but the increases will be small, and it will be some time before they can be translated into concrete capabilities. As armies professionalise, presumably more money will be freed up for defence acquisitions. However, this too is a long-term proposition: France, for example, has found that the professionalisation of its army is costing a lot more than originally expected.

Transatlantic defence cooperation is also a significant factor to consider in weighing the DCI's prospects. While the shortest route to technological compatibility may be off-the-shelf purchases of

U.S. defence wares, such an option is not politically viable among European allies. A better option is defence industrial cooperation between American and European defence industries. However, a major stumbling block is the issue of European access to sensitive U.S. military technology, without which it is difficult to achieve workable transatlantic agreements. On grounds of national security, the United States has established safeguards to protect critical military technology that have effectively limited European access to the U.S. defence market. "The USA's continuing conservative approach to technology co-operation has substantially impeded transatlantic technical co-operation," argues John Hamre, former U.S. deputy secretary of defense.[36] This situation, in turn, is exacerbating the technology gap.

Finally, the EU's decision to create a rapid reaction force will impact DCI implementation. The "headline goal" established at Helsinki has helped increase awareness that military reform is needed, particularly in countries such as Germany, where significant reform is a divisive issue. It has also created a concrete framework for defence reform for those countries that may be motivated by a need to increase EU defence capabilities more than those more broadly of NATO. The nascent reversal in projected European defence spending trends, for example, is being driven not so much by the DCI and NATO as by the EU Rapid Reaction Force initiative. So too are some DCI-relevant undertakings, like the possible short-term lease by France and Germany of American C-17 aircraft, which would be done to meet the EU's 2003 rapid deployment capability deadline.

To date, the EU initiative can be said to be having a positive impact on the DCI. But this will remain true only if measures to strengthen military capabilities in NATO and the EU continue to be consistent with one another. If the EU initiative causes European members of NATO to focus on efforts that do not accord with the DCI, then implementation of the DCI could be stalled or sidelined. The Headline Goal focuses on peace support missions, whereas the DCI captures both lower-level tasks and measures to strengthen NATO's ability to carry out its primary mission – the collective defence of alliance members. Although EU leaders insist that the military requirements, performance levels, and equipment specifications are the same across the two organizations, given the differing mandates, there is potential room for discrepancy as the EU initiative progresses. Both the DCI and the Headline Goal argue that ground forces should have improved C41, sustainability, and strategic mobility, but the DCI places greater emphasis on improvements in "effective engagement" measures like power projection and strike.[37] Already American officials have expressed concerns that in the overall spectrum of conflict, European allies may be concentrating on the "lowest 50 percent and leaving the other (high-intensity) 50 percent to the United States."[38]

One way of bridging this gap may be for European members of NATO to focus on those capabilities that are suited to both peace support missions and high-intensity war. "Whether they focus narrowly on the limited demands of small-scale peacekeeping, or more broadly on the full range of Petersberg tasks, including large-scale combat, will largely determine their ability to reconcile the EU's initiatives with the alliance's DCI."[39] To do so, it is necessary to identify those elements of the DCI, and by extension of the RMA, that are applicable both to the most likely and to the most dangerous tasks NATO could face in the coming decades. It is to this task that the next chapter turns.

CHAPTER SIX

The RMA and Peace Support Operations: A Question of Relevancy

Of the many dimensions the debate surrounding the RMA has taken over the past few years, one of the most important is the question of relevancy. How relevant is the RMA and its associated technologies and doctrines to the types of conflicts Canada and its allies are most likely to face in the foreseeable future? Few would question that the RMA is well suited to high-intensity war against a modern, advanced conventional armed force. But this is only one contingency along the spectrum of possible conflict scenarios, and many would argue it is the least likely in the near to medium term. Rather, most of the conflicts the Western world is likely to address will be low-intensity in nature and involve unconventional forces. The requirement will be less for high-intensity warfare to reverse military aggression between states and more for peace-support operations – from humanitarian assistance to peacekeeping to peace enforcement and post-conflict peace implementation – to take place in response to an intrastate conflict.

This chapter looks at the RMA's applicability to peace-support operations, examining the degree to which key RMA technologies and doctrines were or were not relevant, useful, or effective during recent peace-support operations. Focusing especially on the international experience in the Balkans, the chapter draws conclusions about which areas of the RMA can best be considered as cross-suited to the demands of high-intensity warfare and conflict short of war.

PRECISION FORCE AND
PRECISION-GUIDED MUNITIONS

An early monograph on the military technical revolution argued that joint precision strike forces that coordinated the fire of naval, air, and ground units could be helpful in peace enforcement efforts that called for an outside force to compel action by one or more parties to a conflict.[1] The case was also made that the ability of forces intervening in intrastate conflicts to use precision firepower would likely prove especially useful for minimizing collateral damage and casualties.[2]

At first glance NATO's actions in the Balkans in the 1990s would seem to fully substantiate these views. After the Bosnian Serbs launched a mortar attack on the Sarajevo marketplace at the end of August 1995, NATO began Operation Deliberate Force, which was designed to compel the Serb forces to withdraw their guns from around Sarajevo, shift the military balance in Bosnia toward the Bosnian Croat Federation, and induce the Bosnian Serbs to settle. During the two-week period from 30 August to 14 September NATO flew some 3,400 sorties, involving 750 attack missions against 56 ground targets, such as Bosnian Serb air defences, ammunition depots, artillery sites, and military communication facilities.[3] The precision weapons that NATO used included laser-guided bombs, *Maverick* missiles, and sea-launched *Tomahawk* cruise missiles. A British-French rapid reaction force, which had deployed to the region two months earlier, also launched artillery fire against Serb weapons sites.

Initial reports of air power effectiveness were highly favourable. John White, U.S. deputy defense secretary at the time, said that the strikes were more accurate than those conducted during the Gulf War, with 95 percent of all precision munitions hitting their targets.[4] The Pentagon was more sanguine in its assessment and eventually found that 60 percent of the identified targets were destroyed. Significantly, however, they were destroyed with no collateral damage.[5] A limiting factor was bad weather over Bosnia, which forced aircraft carrying laser-guided bombs to abort about half their bombing missions and contributed to the decision of the alliance to launch satellite-guided cruise missiles.[6]

In the wake of the attacks, Serb forces not only withdrew their guns from around Sarajevo but also agreed to participate in the peace negotiations that culminated in the Dayton Peace Accord of November 1995 and the subsequent Bosnian peace agreement. Such events led to the conventional conclusion that air power, and specifically precision force, was responsible for bringing about the peace in Bosnia. However, a closer look quickly revealed that several other factors were also at work. Most notably, a simultaneous ground offensive by Croatia and the Bosnian Federation, begun some weeks before the air campaign, resulted in the federation and the Serbian republic holding a roughly equal division of Bosnia, making peace much more likely. In addition, the air strikes came at the end of three years of horrible, exhausting fighting and crippling economic sanctions against Serbia. Before the bombing campaign, Bosnian Serb leaders had already agreed to let Serbian president Slobodan Milosevic represent them at peace negotiations. Thus, as one operator has noted, there were simply too many factors in the equation to discern the exact contribution of air power.[7] Nonetheless, it seems clear that air power played an important role in shifting the balance of power on the ground and thus in creating one of the key contextual conditions for negotiating a peace agreement.

A more extensive case study is Operation Allied Force, launched by NATO in and around Kosovo on 24 March 1999. Here the alliance had three objectives in mind: avert a humanitarian catastrophe as a result of Kosovar Albanians being prosecuted by Belgrade; damage Serbia's capacity to wage war against Kosovo; and force Milosevic to agree to a negotiated peace settlement on NATO's terms. During the seventy-eight-day air campaign, NATO aircraft conducted 38,000 sorties, including some 23,300 strike missions against 7,600 targets, of which roughly 3,400 were mobile targets.[8] Approximately 35 percent of the munitions launched were precision-guided.[9] Roughly 60 percent of the target-hit claims made during Operation Allied Force were later confirmed by assessment teams.[10]

NATO used a wide array of precision systems during the campaign. Fighter aircraft from several allied countries, including Canada, conducted tactical air strikes using laser-guided bombs. America's B-2 stealth bombers dropped Joint Direct Attack Munitions (JDAMs) from about 40,000 feet, while their B-52 bombers used conventional air-launched cruise missiles (CALCMs) from standoff positions. NATO ships from the United States and Britain added to this array with *Tomahawk* cruise missiles. Because frequent cloud cover over Yugoslavia made it difficult for pilots to achieve and maintain a laser "lock" on targets, the use of GPS-guided precision systems such as JDAMS, CALCMs, and *Tomahawks* proved particularly critical during the Kosovo operation. In the latter stages of the campaign the Pentagon also deployed the *Enhanced Paveway* warhead, which is guided through the clouds by GPS and to the final target by laser.

The air campaign was of mixed effectiveness. Most analysts would agree that the alliance failed to achieve, at least during the air campaign itself, its humanitarian objective. Indeed, experts have argued that the mass expulsion and victimization of Kosovars was aggravated by NATO's exclusive reliance on air power. That said, in the months following the campaign NATO succeeded in rescuing and resettling over a million refugees. The alliance was also less effective than it had originally hoped and anticipated in degrading the military capability of the Yugoslav forces. Post-conflict damage assessments confirmed that NATO had destroyed 974 mobile targets, including 93 tanks, 153 armoured personnel carriers, 389 artillery pieces, and 339 other military vehicles.[11] The head of the NATO assessment team stated that these figures "amounted to crippling losses for Serbia's regular forces."[12] But they looked much less impressive when compared to the total number of Serbian tanks and armoured personnel carriers in Kosovo – 350 and 440 respectively – before the start of the campaign.

One reason behind the limited military effectiveness of air strikes was bad weather. Determined to keep civilian collateral damage to an absolute minimum, the alliance developed rules of engagement that required visual identification of targets and instructed its pilots to abort those missions that were impeded by cloud cover. As a result, in the first weeks of the campaign almost half of all air combat sorties were unable to attack their assigned targets. In addition, the heavily forested and hilly terrain – so different from what coalition forces faced during the Gulf War – made it difficult for sensor platforms to pinpoint troops and equipment. Following the air campaign it also became clear that sensors had difficulty distinguishing between real and decoy targets. One report stated that NATO had dropped 3,000 precision munitions that resulted in 500 hits on decoys but destroyed only 50 Yugoslav tanks.[13] Meanwhile, the Serb practice of co-locating troops with the civilian population restricted NATO's freedom of maneuver, concerned as it was with limiting civilian casualties. Finally, the allies' concern for pilot safety was a limiting factor in military terms. To ensure NATO aircraft remained out of range of Serbian air defences, allied leaders specified that pilots would not drop below fifteen thousand feet – much higher than the ten thousand feet required for tactical military effectiveness. By contrast, strategic air power was used very effectively to destroy virtually all of Serbia's oil-refining capacity, seriously disrupt its transportation arteries, and cut power to most of Belgrade.

The one area where the alliance achieved its goal was in the broader political objective of forcing Milosevic to agree to a peace settlement. But even here the timing of the settlement, which was signed on 10 June 1999, indicates that although air power played a significant role, other important factors also came to bear. In late May the Kosovo Liberation Army launched ground offensives that forced Serbian units to concentrate and expose their armour and troops, thus making them fully vulnerable to NATO air power for the first time during the war. It is instructive that 80 percent of all Serbian armour losses occurred in the last two and a half weeks of the bombing campaign. At the same time, NATO began to put out signals that it was considering a ground invasion. The effect may have been to convince Milosevic that even if he withstood the intensified air campaign, there was no way out of the conflict short of accepting NATO's terms for a peace settlement. Finally, during this period the Serbian leader faced growing diplomatic isolation, as Russia began to work with NATO to find a peaceful resolution of the conflict. The sharp reduction in Russia's practical support for Serbia was no doubt a significant factor in Milosevic's decision to agree to a peace settlement.

In sum, precision force played a critical role in bringing about an end to hostilities in Bosnia and Kosovo, but in each of these cases it was important for air power to be combined with other initiatives. Moreover, both conflicts were at the higher-intensity end of the range of peace missions. One can expect the impact of coercive airpower to become less decisive as one moves back along the spectrum of conflict from chapter VII peace enforcement to chapter VI peacekeeping and humanitarian assistance operations.[14] Precision air power will also be less effective in less developed countries, such as in Africa, where most civil wars occur, because armoured forces susceptible to air attack are few and far between and conflicts commonly involve roving militias that are difficult to influence through attacks on strategic targets.[15] In such situations it may be more important for an intervening force to be on the ground to negotiate with local commanders, civilian authorities, refugees, or warlords.

Even where a high-intensity response is the solution, the international community will face political and technological limitations to the use of precision force in a peace-support situation. Political leaders will be concerned about limiting civilian and collateral damage. And sensors will continue to have a hard time seeing or tracking weapons and forces hidden within buildings and forests. These limiting factors came together in Bosnia, where surveillance shortfalls and concerns about collateral damage prevented NATO aircraft from effectively countering Serb artillery, mortars, and snipers firing on Sarajevo. It is true that efforts are being made to address these limitations. For example, the United States is experimenting with foliage-penetrating sensor technology, and Britain has identified the need for precision munitions in peace operations that go beyond "highly accurate" to "guaranteeing absolute precision," and therefore eliminating collateral damage.[16] But sensors ultimately face physical principles that simply cannot be overcome, such as being able to "see" inside metal containers. Since precision munitions can only be as accurate as the target information they are acting upon, it is important not to overestimate the present and future value of precision force in responding to intrastate conflicts. Rather, depending on the situation, such force is best seen as a useful tool of first choice for decision makers, which may need to be followed up with other initiatives.

BATTLESPACE AWARENESS AND CONTROL

Many of the advanced intelligence gathering, surveillance, and reconnaissance technologies associated with the RMA and pursued for high-intensity war are highly relevant to peace-support operations. Just as reliable intelligence, surveillance, and reconnaissance (ISR) is central to warfighting, "precise knowledge of how many refugees are moving where, how and under what conditions is critical for effective action."[17] During relief operations in Zaire in the mid-1990s, for example, aid providers were especially in need of data on the number, location, and movement of displaced people.[18] Similarly, to do their jobs properly in Somalia, UN military commanders needed to be able to detect the movement of opposing forces and determine the locations of hidden arms stockpiles.[19]

At the higher end of the conflict spectrum, knowledge of the location of the forces a coalition is trying to target is essential. The "primary challenge in irregular operations is in *identifying* the enemy, not defeating it once it is found."[20] This places a premium on surveillance and intelligence gathering. The relatively high concern to limit allied and civilian casualties in a peace-support mission further raises the importance of advanced sensing technologies. Such systems, for example, were "pressed to their limits" in the Kosovo operation by demands to gather more precise data to protect allied aircraft and ensure that damage and inflicted injury was limited to target areas.[21]

Even for the more traditional, interpositional form of peacekeeping one can envision an important role for the RMA's advanced surveillance technologies. A U.S. Defense Science Board study found that emerging technologies would make possible new standoff approaches to missions that require the separation of combatants.[22] It is conceivable, for example, that in future commanders will be able to separate forces with a "no man's land" populated by remote sensing devices and robotics and enforced with long-range precision strike weapons. This would reduce peacekeeper casualties and improve the chances that a peacekeeping force will remain in theatre long enough for a political resolution of the conflict.[23]

The advanced command and control capabilities that are pursued for war are also highly relevant to peace-support operations. Both types of operations require real-time, integrated communications that link together all military formations. In humanitarian assistance missions enhanced information management technologies are critical to ensuring a coordinated relief effort.[24] In a traditional peacekeeping operation advanced command, control, communications, computers, and intelligence processing (C4I) capabilities, combined with sensing technologies, could allow an intervening force to report to both parties on the deployment of forces on both sides of a cease-fire line in real time, thereby diffusing tensions and assuring both parties of truce compliance.[25] In addition, analysts have argued that improved command and control and communications capabilities are increasingly important for rapidly deploying ground forces that are expected to take on less traditional military tasks.[26] Finally, because both warfighting and peace support operations are increasingly characterized by many nations working together, measures designed to improve multilateral command and control will enhance effectiveness across the spectrum of operations.

Many of these ideas have been reflected in the practical experience of the Balkans over the past several years. In the lead-up to the Bosnian peace agreement in 1995, negotiators used satellite reconnaissance data to provide detail that was unavailable on standard maps and that was crucial for hammering out a consensus on how and where to divide the land between the two entities.[27] Information technology also played a prominent, even decisive, role in convincing the parties that, if signed, the accords would be administered fairly and without prejudice.[28] Since that time, advanced surveillance systems have been used to monitor implementation of the Dayton Accords. In the early stages, Joint Surveillance Target Attack Radar System (JSTARS) ground-surveillance aircraft supervised the exchange of territory between Serbs and Muslims. JSTARS, as well as *Predator* unmanned aerial vehicles (UAVs) and U.S. Navy reconnaissance assets such as the EP-3s, were subsequently committed to long-term peace surveillance operations over Bosnia, monitoring troop movements, illegal arms shipments, arms storage areas, traffic, important government buildings, key bridges, and road intersections. Satellite technology was also instrumental in detecting and publicizing the existence of mass grave sites.

For command and control the U.S. Air Force deployed the Joint Situational Awareness System to integrate information based on returns from JSTARS, UAVs, and Airborne Warning and Control System (AWACS) aircraft. Providing real-time pictures of the battlefield, the system allows commanders to rapidly re-task surveillance platforms to focus on any new hot spots.[29]

A number of advanced surveillance technologies were also at work during the NATO operation in and around Kosovo. JSTARS aircraft located targets and monitored troop concentrations in all weather conditions. AWACS aircraft provided airspace surveillance and directed air-to-air fighters in their operations to provide protection to ground-strike fighters and bombers. Tactical UAVs, like the U.S. Army *Hunter*, operated below cloud-level to provide crucial battlefield reconnaissance on such things as the location of Serbian troops hidden in bunkers or woods. They then sent these images directly back to combat aircraft loitering overhead.[30] Higher altitude U.S. Air Force *Predator* UAVs, which are equipped with synthetic aperture radar and can find targets through cloud cover, augmented the JSTARS. Finally, more than fifty American and European satellites made up of between fifteen and twenty different space-system types were directly involved in NATO intelligence gathering and strike operations.

Despite this array of advanced sensor technology, NATO was not entirely successful in tracking events on the battlefield. When Serbian troops closed the border-crossing points from Kosovo into Macedonia and Albania in early April 1999 NATO reconnaissance planes were unable to locate the tens of thousands of refugees that had been turned back.[31] Sensors also had difficulty locating Serb troops and equipment. One month into the operation, for example, NATO's commanders still seemed unsure exactly where in Kosovo the Serbs' forty thousand troops and four hundred or so armored vehicles actually were.[32] Strikes on fake targets also indicate that the Serbs let NATO daytime reconnaissance flights see real targets and then replaced them at night or that those analyzing sensor information interpreted it incorrectly.

NATO also faced challenges in the area of command and control. Because of the reaction time required to pass data from AWACS aircraft to the command and control centre in Italy and then on to strike assets, the alliance was unable to process information quickly enough to enable aircraft to strike mobile targets.[33] At the same time, NATO experienced interoperability problems, in that allies using older C4I systems could not receive information from America's more technologically advanced systems, such as the secure Joint Tactical Information Distribution System, with which U.S. aircraft are equipped. As a result, U.S. pilots had to rely on voice communications to ensure situational awareness among all allied aircraft. NATO after-action reports stress that these conversations were almost certainly monitored and acted upon by the Yugoslav forces. This thesis was supported in joint testimony by the U.S. secretary of defense, William Cohen, and the chairman of the Joint Chiefs of

Staff, General Henry Shelton, before the Senate Armed Service Committee in October 1999.

Notwithstanding these difficulties, advanced ISR and C4I technologies associated with the RMA gained stature during the Kosovo operation as a result of their performance. America's JSTARS proved so valuable for their ability to track targets through cloud cover that U.S. Air Force officials predict Congress will increase the number of aircraft buys.[34] The older AWACS aircraft demonstrated their worth as a platform capable of coordinating and tracking offensive and defensive air missions, searching for enemy aircraft, assuring safe separation of inbound and exiting aircraft, and directing refueling efforts – all with an extremely high accuracy rate (98 percent during the Kosovo operation).[35] And at a time when keeping allied casualties to a minimum was uppermost in leaders' minds, UAVs provided critical battlefield information that would otherwise have had to have been gathered by low-flying aircraft (aircraft that were therefore vulnerable to enemy fire).

These facts lend credence to the view that advanced surveillance and command and control technologies associated with the RMA can play a key role in the success of peace support operations. Moreover, continued advances in ISR and C4I capabilities are likely to benefit warfighting and peace support missions almost equally.[36] More sensitive heat and motion detectors, the ability to look more reliably through clouds and jungle, and the ability to define the battle down to small groups of soldiers are all high-technology surveillance capabilities that are being pursued for more effective warfighting but are very relevant to peace missions. Similarly, trends in airpower technology indicate that certain command and control shortfalls, such as real-time communications allowing for strikes against mobile targets, are likely to be eventually overcome.[37]

FORCE PROJECTION AND STEALTH

One of the key elements of the RMA is increasing force projection capabilities. In part this comes from making ground forces lighter and more rapidly mobile, but it also involves technological advances that enable air force platforms to travel further and longer and to have more room for maneuver once they are in the battle area. The critical technological advance in this latter area is in low-observable technologies, or stealth.

The Kosovo operation reaffirmed the lesson of the Gulf War that the American F-117A stealth fighter is a highly effective platform for carrying out missions where nonstealthy aircraft would be placed at undue risk of being hit by enemy antiaircraft fire. This is the case despite the fact that a stealth fighter was shot down early in the campaign. The real "surprise" of the Kosovo air campaign, however, was the proven accuracy and reliability of the B-2 stealth bomber. The U.S. Air Force had long boasted that the B-2 gave it "global reach, global power," but this was the first time it had demonstrated such capability in a sustained operation.[38] Flying thirty-one-hour round trips from Whitman Air Force Base, Missouri, the B-2s carried out thirty-three missions, hitting with 90 percent accuracy targets such as Serbia's integrated air defence system, command and control sites, runways and airfields, communications facilities, factories, bridges, and other elements of infrastructure.[39] In the aftermath of the war General Wesley Clark, supreme allied commander Europe at the time, identified continued reliance on stealth aircraft as a key lesson of the Kosovo operation.

Others go further and argue that force projection should in future increasingly focus on stealthy, unmanned strategic platforms. NATO's experience in Kosovo demonstrated both that it is difficult to fight an air war from above the clouds and that allied leaders may be hesitant to send manned aircraft below the clouds. Logically speaking, then, these trends are likely to drive air forces towards unmanned combat aerial vehicles. Along these lines, a recent National Defense University Strategic Assessment argues that the F-22 and Joint Strike Fighter may be the last low-flying tactical combat aircraft purchased by the United States.[40]

But while stealthy platforms may be useful, and even central, to the effective application of force in a higher-intensity peace support mission, it is difficult to discern a role for them in missions at the lower end of the conflict spectrum. Certainly, interpositional peace-keeping missions with cease-fire lines that are monitored by remote sensors and robotics and enforced with standoff munitions could benefit from the presence of unmanned combat aerial vehicles that can hover on station for many days at a time. Indeed, having a sensor and strike capability integrated into the same unmanned stealthy platform could reduce tensions by increasing the credibility of the idea in the eyes of the parties that a violation of the zone of separation would be met with retribution.

Beyond this, however, there is much to be said for the human contacts that are at the root of effective peacekeeping and humanitarian assistance missions. "Close contact is the sine qua non of armies, and it gives them unequalled ability to come to grips with local conditions, distinguish between allies and enemies, and execute schemes to shape social and political developments."[41] Regardless of the technology available, the effectiveness of a monitoring and patrolling mission may be highly dependent on the establishment and maintenance of human intelligence networks. And when it comes to humanitarian missions against units conducting ethnic cleansing with small and medium weapons – which are easy to hide inside buildings and vehicles and therefore remain essentially undetectable by sensors – those units could still operate effectively unless challenged by a ground force of comparable strength.[42] This being the case, it is useful to examine the relevance of RMA-related land force developments to peace support operations.

SMALLER, MORE RAPIDLY MOBILE, AND FLEXIBLE GROUND FORCES

The doctrinal change associated with the RMA that calls for a move towards more rapidly mobile and flexible ground forces that are still highly lethal and can operate in a "nonlinear" environment resonates well with the requirements of future peace support operations. Indeed, the smaller, more highly skilled professional army units that will be called upon to fight any future high-intensity war are likely to be well suited to the complexity of tomorrow's peace missions. The associated organizational trend towards modularity and task-tailored forces lends itself well both to peace operations and to major theater war. However, smaller-scale contingencies place especially heavy demands on combat support (such as construction engineering) and combat service support (like logistics and medical) units. A focus entirely on warfighting would not likely address these requirements; modest force structure changes would be able to accommodate them.[43]

Equipment trends associated with the RMA's ground force doctrine are relevant to peace support operations. Smaller army platforms that still afford troops significant protection and – armed with precision munitions – remain highly lethal, resonate well with the nature of today's peace missions, in which peacekeepers are often greeted with narrow or nonexistent roads and infrastructure and a volatile and dangerous operating environment. America's Future Combat System of vehicles and Britain's Future Rapid Effect System are particularly relevant here. The tanks they are to replace, the *Abrams* and *Challenger 2*, respectively, have proven too unwieldy in places like Kosovo.[44] But a platform with the tank's capability is still needed: "In the end, armor provides the same protection against a rocket-propelled grenade, whether it is fired by regular forces in war or by irregular forces against peacekeepers."[45] The capture of UN peacekeepers by rebel forces in Sierra Leone demonstrated that military units, though deployed for peace operations, must be prepared for hostilities up to and including combat.[46] So, too, did the U.S. experience in Somalia in 1993 and that of the UN in Bosnia in 1992 to 1995.

Whether a force is responding to a high-intensity war scenario or a peace support situation, political leaders will want it to be rapidly deployable to the battle area. Indeed, one of the key lessons of the Balkans is that the general-purpose forces that are most suited to peacekeeping must be rapidly deployable. During the Kosovo operation it took several weeks longer than expected to deploy Task Force Hawk to Albania, because of the weight of the force. Experts have argued that technologies that can make intervention forces lighter will permit them to be deployed quickly to stop genocides or other low-level, yet severe, forms of violence.[47] Strategic sea and airlift assets, which are central to the effective prosecution of a high-intensity war, are also key to rapid deployability in a peace support context. In the aftermath of its 1999 operation in East Timor, Australia identified not only the need for rapidly deployable ground forces but also the imperative of strategic sea and air lift assets to respond to such situations.[48]

The RMA also calls for the increased use of transport and combat helicopters for mobility on the "battlefield." Here, too, there is applicability to peace support operations, particularly with respect to

transport helicopters, since poor road conditions often make helicopters the only viable means of moving within the theatre. Following massive flooding in Mozambique in early 2000, the only means
for the international community to transport supplies into the inte-
rior of the country was by transport helicopter.[49]

Based on the Kosovo experience, the utility of combat helicopters
in a peace mission is less clear-cut. Soon after the air campaign began, General Clark requested the deployment of *Apache* combat helicopters because they were better suited than higher-flying strike
aircraft to degrading the Yugoslav ground forces. But the u.s. political leadership hesitated to do so for fear that the low-flying, slow-
moving helicopters would be vulnerable to automatic-weapon
ground fire or short-range shoulder-fired missiles. In a campaign
where the United States strove towards a "zero casualty" goal, such
considerations played a key role in decision making. Although
President Clinton eventually ordered two dozen *Apaches* and their
associated support personnel and equipment to Albania, in the end
they were never used in combat.

The ultimate decision not to use combat helicopters in Kosovo,
coupled with NATO's more general reluctance to commit ground
forces to the operation, raises questions about the role of the army
in future peace support operations. As noted above, ground forces
are indispensable for certain kinds of peace mission, like traditional
peacekeeping and humanitarian missions. They are also central to
efforts at peace implementation after a settlement has been reached.
In contrast to the Kosovo experience, for example, army tactical
aviation was used extensively in Bosnia after the peace accord was
signed to monitor whether the sides had removed their warfighting
equipment from the zone of separation and to threaten retribution
if compliance was not immediately forthcoming.[50]

For the higher-intensity form of peace support mission, that is to
say peace enforcement, the role of the army is less certain. The crux
of the problem lies in the nexus of interests, risks, and the degree to
which leaders are willing to sustain military casualties. Broadly
speaking, high-intensity war responds to a threat to vital interests,
while peace support missions more often than not address core values – things that offend our conscience and fuel our outrage but do
not, in the final analysis, threaten our livelihood or way of life. This
low threat to national interests translates into a low tolerance for

casualties on the part of publics and political leaders. But the fact that a mission may be responding to a threat to values does not reduce the risk to forces. Indeed, "peace operations are distinguished from open conflict not by the types of tactical operations undertaken but by their intent."[51] While the objective of the use of force in war is the reversal of aggression, its goal in a peace support operation is to halt hostilities as soon as possible, as a precursor to bringing about peace. It follows that a peace enforcement mission can contain the same degree of risk to forces as does a warfighting mission, but in a situation that does not threaten vital interests. Since ground forces are, generally speaking, at greater risk of sustaining casualties than the other services, political leaders are likely to hesitate to employ them in a peace enforcement operation.

JOINTNESS AND LITTORAL WARFARE

The RMA's trend towards littoral warfare goes hand in hand with that of jointness, in that the conceptual move towards naval power-projection from the sea onto land necessarily requires naval forces to work in concert with ground forces, and possibly air forces as well. Early on in the RMA debate analysts made the case that when responding to "irregular operations," navies would have to become more capable of operating close to shore and with other services in the context of littoral warfare.[52] Recent peace enforcement operations in Kosovo and East Timor provided some confirmation of these trends and, by extension, of the relevance of these concepts to peace missions. During the NATO air campaign, a significant portion of the precision strikes were carried out using *Tomahawk* cruise missiles launched from U.S. and British ships in the Adriatic littorals. Indeed, cloudy weather rendered the GPS-guided *Tomahawk* a critical asset in the alliance's arsenal. This was similarly the case during the much shorter Bosnian mission four years earlier. In East Timor, the Royal Australian Navy's helicopter support of land forces that were deployed ashore also drew light on the valuable contribution that joint littoral operations can make to the effective conduct of a peace enforcement mission.[53]

Beyond this, jointness was also exhibited in an air-land context in Kosovo. In the early stages of the mission, military planners broadened the operational plan for JSTARS from supporting only the air campaign to assisting army ground and helicopter units. Had the *Apaches* been deployed to Kosovo, the two platforms would have worked closely together, with the surveillance aircraft orbiting before the launching of the attack helicopters, in order to provide them with an updated view of Yugoslav forces. In addition, the Kosovo Liberation Army ground offensive in late May 1999, which increased air power effectiveness by forcing Serb units out into the open, demonstrated the importance of joint air-ground coordination.

CONCLUSION

Thus, an analysis of recent international missions reveals a relatively high degree of relevance of key RMA technologies and doctrines to peace support operations. Precision force and precision guided munitions can be useful in a peace enforcement mission and may also have some application in an interpositional peacekeeping mission. But the international community will often face political and technological constraints in their use. For this reason, standoff precision force is best seen as an important tool of first recourse in peace enforcement and one that increases decision makers' options. Advanced surveillance and command and control technologies are very relevant to the entire range of peace support operations, whether the task is humanitarian assistance, peacekeeping, peace enforcement, or peace implementation. Yet although trends in technology indicate that today's command and control shortfalls will eventually be overcome, there will always be certain sensor limitations when it comes to intelligence, surveillance, and reconnaissance. Stealthy strategic platforms proved highly useful, even crucial, during the Kosovo peace enforcement operation. However, they would have little or no application to other types of peace mission. The role of the army in future peace enforcement operations is unclear because of the high-risk–low-interest nexus inherent in these missions. That said, ground forces remain central to the other forms of peace operation. Moreover, RMA measures to make armies smaller, lighter, more rapidly mobile and deployable, and yet still highly lethal are directly in line with the ground force requirements of tomorrow's peace support missions. Finally, the RMA doctrines of jointness and littoral warfare often have direct application to peace operations.

In short, there are certain aspects of the RMA, such as stealthy bombers, that have only limited relevance to peace support operations or that will only be useful at the high end of the spectrum of these missions. There are also elements central to peace support operations, such as combat service support units, that will be in insufficient supply if the focus is placed solely on force attributes for warfighting. But, generally speaking, the contrast that is often drawn between the high-tech, high-intensity requirements for war, and the low-tech, low-intensity means for peacekeeping is an artificial one. Better technology can help compensate for the political dependence on nearly casualty-free operations that is especially present when core interests are not at stake. The relatively high degree of relevance of RMA technologies and doctrines to peace support operations indicates that if a nation prepares well for war, it will also be well prepared for peace.

CHAPTER SEVEN

Asymmetric Threats

The revolution in military affairs brings with it the possibility, perhaps even likelihood, that America and its allies will find it increasingly difficult to operate together. But there are other, less direct implications of the RMA that are difficult to identify and still harder to quantify but that are nonetheless very real. America's pursuit of the RMA is affecting not only its allies but also its potential enemies. Unable to match its sophisticated technologies, adversaries are seeking to gain advantage over the United States by using *asymmetric* means to undermine U.S. strengths while exploiting its vulnerabilities. The U.S. Joint Chiefs of Staff have described asymmetric approaches as attempts to circumvent or undermine U.S. strengths while exploiting U.S. weaknesses, using methods that differ significantly from the United States' expected mode of operations.[1] Such methods could include the use of weapons of mass destruction (WMD), that is, nuclear, biological, or chemical weapons, as well as terrorism, information warfare, or limited ballistic missile attack.

The concept of asymmetric threats is closely linked to that of *homeland defence*, which refers to measures to defend the people, property, and systems of the United States from a variety of nontraditional, asymmetric threats. Although U.S. forces and facilities overseas face asymmetric threats, for several years U.S. defence analysts have believed that a key area of U.S. weakness lies in its ability to protect its homeland. An attack on the U.S. homeland, such as that carried out on 11 September 2001, represents the most extreme expression of an asymmetric strategic attack.[2]

This chapter examines the nature of the asymmetric threat to North America. It discusses various aspects of the threat, measures both the United States and Canada are taking to respond to them, and how the increased asymmetric threat to the u.s. homeland is likely to affect Canada-u.s. relations. It finds that America's pursuit of the RMA could well encourage adversaries to target the United States "asymmetrically" at home again and that by virtue of its geographical connection, Canada is likely not only to be directly affected but also to be drawn into American responses. In the coming years, the well-being of Canada's vital security relationship with the United States will depend significantly on the extent to which it supports America's continental priorities.

THE NATURE OF THE ASYMMETRIC THREAT

Since the mid-1990s the United States has increasingly focused on the concept of asymmetric warfare. The term does not appear in the u.s. military's 1990 Base Force, 1993 Bottom-Up Review, or 1995 Commission on Roles and Missions of the Armed Forces. But in 1996, *Joint Vision 2010* referred directly to the risk of "asymmetrical counters to u.s. military strengths, including information technologies." The 1997 Quadrennial Defense Review expanded on this theme, arguing that u.s. dominance in the conventional military arena may encourage adversaries to use asymmetric means – including ballistic missiles, weapons of mass destruction, terrorism and information warfare – to attack forces and interests overseas and Americans at home. Both the October 1998 and December 1999 versions of *A National Security Strategy for a New Century*, released by the White House's National Security Council, discussed asymmetric warfare, as have all of the annual reports to Congress by the secretary of defense since 1998.

In recent years experts outside government have also stressed the asymmetric threat. The 1997 report of the congressionally mandated National Defense Panel, which paralleled the QDR effort, stated pointedly that "We can assume that our enemies and our adversaries have learned from the Gulf War. They are unlikely to confront us conventionally ... [Rather] they will look for ways to match their strengths against our weaknesses." It went on to place Homeland Defense as America's number one national security challenge in the period to 2020. The 1999 report of the u.s. Commission on National Security/Twenty-first Century went still further, concluding that American interests overseas would face a growing range of threats and stating with foresight that the United States would become increasingly vulnerable to hostile attack on its homeland.

These analyses have been based on a confluence of incentives and means. In the first instance, the dramatic increase in America's conventional capabilities has made it unlikely that future adversaries will confront the United States on the traditional battlefield. "Asymmetrical assault is the only strategy that makes sense for erstwhile enemies," argues one analyst, "since they cannot compete under the traditional rules."[3] At the same time, it has become easier for these same adversaries to find unconventional means of attack. Since the end of the Cold War, weapons of mass destruction, including nuclear, biological, and chemical weapons, have proliferated significantly. The number of actors with access to nuclear weapons has gradually expanded beyond the initial five nuclear weapons states to encompass numerous players who are not bound by international treaties and obligations restricting the development or use of such capabilities. They include "states of concern" such as Iran and North Korea; regional powers like India and Pakistan; religious, ethnic, and nationalist groups; terrorists; and, possibly, criminal organizations.[4]

Biological and chemical weapons are even more readily accessible to terrorists. According to intelligence sources in Europe and the United States, militant political groups "across the globe" are developing or seeking to purchase biological weapons for terrorist use.[5] Statistics indicate that of the seventy or so terrorist incidents involving chemical or biological elements world-wide in the twentieth century, well over half occurred in its last decade.[6] In the weeks following the terrorist attacks on New York and Washington, CIA officials stated that Osama bin Laden's global terrorist network, Al Qaeda, had been trying for some time to acquire chemical and biological weapons.[7] In addition, CIA director George Tenet has testified that some twenty countries have or are actively developing chemical or biological weapons, among them several countries that are openly hostile to the United States.[8]

The means of delivering weapons of mass destruction have also increased. No longer limited to traditional methods, such as bombers and sophisticated ballistic missiles, highly destructive devices can now be transported in small trucks or cargo containers.[9] Many of the tools of asymmetric warfare may now be well within the reach of less-developed countries that could not begin to challenge the United States with "traditional" military approaches.[10] Indeed, some military experts have argued that asymmetric capabilities could be the "great equalisers" of the twenty-first century as the great powers' sources of strength become the smaller countries' points of attack.[11]

As for ballistic missile technology, in 1998 a bipartisan panel of experts, the Rumsfeld Commission, concluded that new ballistic missile systems being developed by North Korea and Iran would be able to inflict "major destruction" on the United States within five years of a decision to acquire such a capability. Moreover, during several of those years the u.s. might not be aware that such a decision had been taken. This is because the ability of intelligence agencies to monitor the emerging threat is eroding as nations become increasingly sophisticated at concealing evidence of ballistic missile activity and have greater access to technical assistance from outside sources. In 1999 the CIA concluded that any country, regardless of its missile development experience, could field an intercontinental ballistic missile by 2015.[12]

Information warfare is also increasingly an option for those who would seek to challenge u.s. power with asymmetric means. This term is defined by the Office of the Chairman of the u.s. Joint Chiefs of Staff as actions taken to achieve information superiority by affecting adversary information, information-based processes, information systems, and computer-based networks, while defending one's own information, information-based processes, information systems, and computer-based networks.[13] One form of information warfare is computer network attack, or "hacker attack," which refers to software-based attacks on information systems. Such attacks directly corrupt information without visibly changing the physical entity within which it resides. As a result, a computer network attack can take place without the targeted entity realising it.

The growing dependence of modern societies and their military forces on computers and computer networks has created points of vulnerability that may be easily exploited by computer hackers. Potential attackers range from national intelligence and military organizations, terrorists, and criminals to industrial competitors, hackers, and disgruntled or disloyal insiders. At least a dozen countries, some hostile to America, are reported to be developing an information warfare capability.[14] Information warfare may be particularly attractive as a tool of terrorism because it gives individuals and groups a reach that was previously reserved for well-organised, state-funded terrorist organizations. "Physical distance and national borders that once separated terrorists from their co-conspirators, their audience and their targets cease to exist in the world of modern telecommunications and the internet."[15] Indeed, the reduction in strategic depth that began with the airplane and was accelerated by the military and commercial use of space has all but collapsed with the explosion of information technologies, negating many traditional concepts of state security.[16]

Finally, it goes without saying that the tragic events of 11 September 2001 have dramatically highlighted the fact that, notwithstanding the increased prospect of WMD or cyber warfare, conventional terrorist threats remain ever present. "This goes to prove the whole argument you don't need weapons of mass destruction," argued one WMD expert soon after the attacks; "all you need is an airliner loaded with jet fuel."[17] It has long been evident that terrorists motivated to inflict mass casualties can do so using traditional means, which generally pose fewer technical difficulties than do WMD attacks. Indeed, bombs have been responsible for over three-quarters of the terrorist incidents that have killed 100 people or more over the past quarter century.[18]

ASYMMETRIC THREATS IN THEATRE

Pentagon analysts foresee several asymmetric threats that could affect the conduct of future operations in the theatre of war. Weapons of mass destruction could be used to deny U.S. and allied forces access to key facilities (such as ports and airfields), deter allies and potential coalition partners from supporting U.S. intervention, or inflict higher-than-expected U.S. casualties in an attempt to weaken national resolve. Traditional terrorist methods could be used against American forces and interests, as illustrated by the bombing of the U.S. barracks in Saudi Arabia in 1996 and the U.S. embassies in Africa in 1998. Information warfare tactics against the United States and its allies in theatre may be a particularly attractive option, given the inherent vulnerabilities associated with relying increasingly on information technology. Enemies could block access to intelligence information through dedicated electronic warfare or computer network attacks on manned, unmanned, and satellite-based surveillance and command and control systems. Already U.S. war games have indicated that by 2025 or earlier the U.S. space systems upon which much of the U.S. military's information depends could be at risk from information warfare attack.[19]

To address these threats, the U.S. military has taken several steps. Among them, the air force has developed an "asymmetric vulnerability assessment tool" to help predict what would happen if an enemy shut down ports and airfields with chemical weapons or other means of unconventional attack. The Joint Special Operations Command has established covert action teams, known as Special Mission Units, that are specifically manned, equipped, and trained to counter terrorist use of weapons of mass destruction in theatre. And the Pentagon has vaccinated almost every member of the U.S. armed services against anthrax, one of the deadliest biological weapons, to protect them against a biological-weapon attack in theatre. To guard against information warfare attack, military planners are installing protective weapons on satellites, shielding their internal electronics, and improving satellite capabilities to function in an electronic jamming situation. Beyond this, the U.S. Army has established a Computer Emergency Response Team to respond to the service's computer security problems around the world, as has the U.S. Navy. The U.S. Air Force has created an entire Information Warfare Squadron to provide a deployable counter-information warfare capability at the disposal of theatre commanders.

ASYMMETRIC THREATS AGAINST THE UNITED STATES AT HOME

Since the March 1995 sarin gas attack on the Tokyo subway by the Japanese Aum Shinrikyo cult, fears of a terrorist event involving weapons of mass destruction taking place on the territory of the United States have grown substantially. Although the United States has not experienced any mass-casualty incidents involving WMD terrorism on its homeland, individuals associated with domestic extremist organizations have been arrested for possession of chemical and biological agents, and WMD-related scares (subsequently determined to be hoaxes) have prompted large-scale emergency responses in several U.S. cities. In light of these incidents, senior U.S. national security officials have publicly described WMD terrorism as one of the most serious threats facing the United States. In 1999 former U.S. president Clinton stated it was "highly likely" that a terrorist group would launch or threaten a germ or chemical attack on American soil within the next few years.[20] Following the terrorist attacks of 11 September 2001 the U.S. Federal Aviation Authority grounded crop dusters as a result of a "serious credible threat" that such aircraft could be used to conduct chemical warfare.[21]

At the same time, many military and intelligence officials in the United States believe that a computer network attack on infrastructures at home is a "real and growing" threat.[22] Such infrastructures may include those of transportation, oil and gas production and storage, water supply, emergency services, banking and finance, electrical power, information and communications, and government services. Already there is evidence that government and private industry systems in the United States and Canada are "constantly under systematic and organised" information attack.[23] In 1999 the Pentagon reported that it was subjected to daily information attack, with between eighty and one hundred "cyber incidents" on its computer systems each day.[24] This figure doubled in 2000.[25]

While military systems are at risk from computer attack, they are not nearly as vulnerable as the civilian computer systems that are critical to the smooth functioning of modern society. A three-year study by the Washington-based Center for Strategic and International Studies released in 1999 found that more than twenty countries had already penetrated U.S. systems, with most cases being industrial espionage.[26] Another study argued that organised criminals, terrorists, and foreign spies are likely to be launching widespread information attacks on vital U.S. infrastructures by 2005.[27] While redundancies are inherent in a networked environment, the "explosive growth" in computer connectivity over the past ten years has significantly increased the risk that vulnerabilities exploited within one system will come to affect another system.[28]

It is not difficult to imagine the dire consequences for civil society of severe disruptions in infrastructures ranging from electricity grids and telephone services to air traffic control and water purification systems. In addition, a computer network attack on civilian systems could have direct military implications. American defence experts have noted that nearly everything the u.s. military does – from designing weapons systems and guiding missiles to paying, training, equipping, and mobilising soldiers – depends on computer-driven civilian information networks. Over 95 percent of u.s. military communications during peacetime travel over the same phone lines as those that are used by civilians.[29] The national electric grid powers American military bases, Pentagon purchases are conducted via the federal banking network, and soldiers are transported on aircraft directed by civilian air traffic control towers.[30] In short, "if the civilian computers stopped working, America's armed forces couldn't eat, talk, move or shoot."[31] The Canadian forces face a similar set of dependencies and challenges.

Finally, the attacks on the World Trade Centre and the Pentagon have brought to the fore the longstanding conventional terrorist threat to the u.s. homeland and the physical dangers faced by key u.s. infrastructures. In anticipation of a "second wave of terror," deemed likely by high-level u.s. officials,[32] the Bush administration has increased its focus on the traditional terrorist threat to the American homeland, even as it continues to assess the threat of other forms of asymmetric warfare.

RESPONSE

The United States has taken several steps to respond to asymmetric threats to the homeland. The Federal Bureau of Investigation (FBI) and the Federal Emergency Management Agency (FEMA) already share lead-agency status for WMD-related terrorist incidents in the United States. In the event of such an incident, DOD assets would be integrated into a coordinated federal response effort led by the FBI, for crisis response, and by FEMA, for "consequence management," that is, managing the fallout from a WMD incident. Within the DOD, the army is the lead service for planning the domestic defence against and recovery from terrorist use of weapons of mass destruction. The DOD currently maintains forces that can be tasked on a twenty-four-hour on-call basis to assist in responding to a WMD-related terrorist incident. In addition, as part of its Domestic

Preparedness Program, in 120 U.S. cities the DOD is helping to train "first responders" – local police forces, fire departments, hazardous material specialists, and emergency medical personnel – to deal with the terrorist use of WMD. The United States has also created a permanent task force within Joint Forces Command to coordinate the military's response to a chemical or biological attack on the United States.

In 1999, in response to the prospect of a rogue state launching a ballistic missile attack, President Clinton signed the National Missile Defense (NMD) Act, calling for the implementation of a system to protect against limited attacks as soon as technologically possible. Influenced by the perception that such a threat could materialise sooner rather than later, Congress passed the act by an overwhelming majority. In the fall of 2000 Clinton deferred a decision on whether to proceed with NMD deployment for the next administration. The Bush administration has stated its intention to build and deploy a (now-renamed) ballistic missile defence system and is in the process of conducting consultations with NATO and Russia as a means of doing so in cooperation with friends and allies. At the G-8 summit in Genoa in July 2001, Bush and Russian president Vladimir Putin agreed to link discussions of American plans to deploy a ballistic missile system to the prospect of large cuts in the U.S. and Russian strategic arsenals. Following the terrorist attacks on the United States, where the seemingly impossible became the possible, any lingering congressional opposition to the missile shield has all but evaporated.

America has also taken steps to respond to the threat of cyber terrorism. In 1996 Clinton created the President's Commission on Critical Infrastructure Protection to examine possible threats to critical infrastructures. In its 1997 report, *Critical Foundations: Protecting America's Infrastructures*, the commission recommended the creation of a national organizational structure to address the cyber threat. This included sector coordinators to promote cooperation within industry, lead agencies to serve as conduits from government to each sector, the Infrastructure Sharing and Analysis Center to establish the magnitude of the threat, and the high-level Office of National Infrastructure Assurance, associated with the National Security Council. Presidential decision directive number 63 of 1998 implemented the recommendations. Also in 1998 the

FBI created the National Infrastructure Protection Center, charged with being the federal government's focal point for gathering information about threats and coordinating any response to incidents affecting key infrastructures. In 2000 the White House released its *National Plan for Information Systems Protection* as the first element of a more comprehensive effort to protect the nation's information systems and critical infrastructures from future attack.

Meanwhile, in 1998 the Pentagon established a Joint Task Force for Computer Network Defense to oversee and coordinate efforts to protect DOD computer systems from cyber attack. These efforts involve coordinating actions among the various computer emergency response teams and the wider U.S. intelligence and law-enforcement communities. In October 1999, U.S. Space Command assumed responsibility for defending all defence department computer networks from hacker or foreign attack.

Despite these measures, high-level American policymakers and defence experts have argued that the U.S. government has identified only the "tip of a very large iceberg" and that the United States is still failing to adequately protect vital computer networks against a crippling cyber attack at the strategic level.[33] "Strategic information warfare" refers to a coordinated information warfare attack that involves such things as extended terrorist campaigns; attacks by organised crime; industrial espionage; and organised, strategically targeted information-technology operations undertaken for the purposes of defeating an enemy. This kind of warfare is distinct from hacker attacks, which are usually conducted "either for thrills or to send a message, but not over a long period of time for the purpose of winning a war."[34]

Some critics have argued that the government must undertake greater cooperation with the private sector, which owns and operates many of the computer networks upon which defence and other government departments depend. Others have emphasised a need for an overarching command or agency to coordinate the government's response to WMD and cyber threats. Such was the conclusion of the third report of the United States Commission on National Security/ Twenty-first Century, released early in 2001. Arguing presciently that a direct attack against American citizens on American soil was likely over the next quarter century, the commission recommended the creation of a new independent National Homeland Security

Agency with responsibility for planning, coordinating, and integrating U.S. government activities involved in homeland security.

Responding to these concerns, and following the attacks of 11 September 2001, President Bush announced the creation of a new Cabinet-level Office of Homeland Security, to which both an Office of Cyberspace Security and an Office for Combatting Terrorism will report directly.[35] The overall office will integrate the homeland security responsibilities of more than forty government agencies. Beyond this, it will likely define the military's role – which could be substantial, given the defence resources available in the reserves and the National Guard. Such a role is to be clearly subordinated to law enforcement and civilian authorities.[36]

Some have recommended that the U.S. military combine all its missions related to the defence of the United States and to support for civil authorities into a single U.S. command. But a report on the issue released in early 2001 argued against this move in favour of continuing the trend towards consolidating missions related to the defence of the United States in Space Command and those related to support to civil authorities within Joint Forces Command.[37] In the wake of the terrorist attacks on America, however, the Pentagon is re-examining how it can better integrate its homeland defence responsibilities.[38]

CANADA AND ASYMMETRIC WARFARE

A recent study of the asymmetric threat commissioned by the Department of National Defence states that the risk of a large-scale asymmetric attack on Canadians on Canadian soil is slight. Even in the aftermath of the attacks on the United States, the prime minister played down the risk of any terrorist campaign spreading into Canada. More likely, Canada would be an indirect target of asymmetric warfare. Because of our interconnectedness with the United States, any attack against the United States, particularly in the realm of information operations but also conceivably with respect to WMD terrorism, would significantly affect Canada. Canada is also considered a likely launching ground for asymmetric attacks against the United States – witness the Ahmed Ressam case, in which a Canadian-based terrorist planned to conduct an attack on the Los Angeles airport on the eve of the year 2000. Finally, the fact that the effect of even a single asymmetric attack on Canadian soil, however remote, could be of such catastrophic significance to Canadian security has warranted an increased focus on such threats.

Asymmetric threats to Canada and Canadians overseas have figured increasingly in recent policy documents. Within the Department of National Defence (DND), the *Strategy 2020* document of 1999 provides strategic direction for the Canadian forces to "deliver a joint capability to deal with weapons of mass destruction, information operations, and other asymmetric threats." It also sets a five-year modernization target for developing "new task tailored capabilities to deal with asymmetric threats and weapons of mass destruction." *Defence Planning Guidance 2001* tasks the department to develop an implementation plan for these capabilities by mid-2002.

The DND is the lead agency for addressing asymmetric threats when the threat is to DND assets and Canadian forces operations. Government-wide, the *National Counter-Terrorism Plan* identifies the solicitor general of Canada as the lead minister for the National Counter-Terrorism Program and for coordinating Canada's response to terrorist incidents within Canada, including WMD terrorism and physical attacks on critical infrastructure. In the event of such an incident the Royal Canadian Mounted Police (RCMP) would head up law-enforcement aspects, while Emergency Preparedness Canada (EPC) would coordinate "consequence-management" efforts. This division of responsibilities echoes that between the FBI and FEMA. The DND's role would be to assist the RCMP and the EPC in threat identification, crisis response, and consequence management. To this end, the Canadian forces have two dedicated crisis-response organizations for counter-terrorism, Joint Task Force 2 and the Nuclear, Biological, and Chemical Response Team.

To address the threat of information warfare attacks against Canada's critical infrastructures, the federal government has recently created an Office of Critical Infrastructure Protection and Emergency Preparedness. Although it has been established within the DND, the office has responsibilities that cut across traditional departmental lines of authority and coordinates many of the intelligence and counter-terrorism activities of the RCMP, the Canadian Security Intelligence Service, and the Communications Security Establishment. To the extent that critical infrastructures are also at risk of a physical attack, such as the bombing of an electrical facility, the EPC is also involved, and these civil defence functions have also been integrated into the Office of Critical Infrastructure Protection and Emergency Preparedness.[39]

Following the terrorist attacks on America the Canadian government announced a series of anti-terrorism measures, including the creation of a powerful cabinet-level committee on security, roughly equivalent to the U.S. Office of Homeland Security.[40] Led by Foreign Affairs Minister John Manley, and including all the cabinet ministers in charge of departments directly involved in the response to terrorism, the committee is charged with developing a tough anti-terrorism package and coordinating Canada's actions with Washington.[41]

For the DND itself, the Information Operations Group is mandated to address the threat of cyber attacks against defence establishments, assets, and resources – a role similar to that of Space Command in the United States. It is supported by a team of scientists that is developing techniques for attack detection and analysis, information protection and assurance, and information exploitation.[42] DND has also set up a Critical Infrastructure Protection Working Group, and the Canadian forces have established both a Network Vulnerability Assessment Team and a Computer Incidents Response Team. Finally, efforts to coordinate the defensive information operations of the American and Canadian militaries are taking place within NORAD and, more generally, under the direction of the Canada-U.S. Military Cooperation Committee.

IMPLICATIONS FOR CANADA

America's increasing focus on asymmetric threats and particularly the homeland defence dimension holds significant implications for Canada. Even before the terrorist attacks on New York and Washington, its concern about such threats was directly affecting Canada-U.S. defence relations. Over the past few years the ballistic missile debate had evolved in the United States such that it no longer centred on *whether* but on *when* to deploy a limited ballistic missile defence system. Although America can field such a system with or without Canadian involvement, its clear preference is for a bilateral deployment. Canada can expect growing pressure to participate and may ultimately be faced with the decision of whether to do so despite its concern about the impact of such a system on the existing arms control regime, or whether to accept a situation in which the United States conducts surveillance over and defends Canadian territory without Canada's say in the matter. In addition, U.S. Space Command's recently added responsibility for preparing U.S. forces to conduct cyber attacks against enemy computer networks could increase pressure on Canada to adopt a similar approach.

Beyond this, other aspects of America's homeland defence agenda will demand Canadian involvement. Canada shares an integrated geographic and economic space with the United States, with tightly linked networks for road, rail, and water transportation; air traffic control; information; oil and gas delivery; hydro electricity distribution; and finances. America simply cannot address these issues with the degree of effectiveness it wishes to have without participation by Canada. The United States would like Canada's cooperation in protecting critical infrastructures, guarding against weapons of mass destruction, and taking measures to counter other security problems associated with an open and porous border. Whether it be a cyber attack against telephone lines, a WMD threat to water systems, or explosives driven across the border, the degree of interconnectedness between the two countries is such that neither country can fully address domestic asymmetric challenges without cooperation by the other. The notion of Canada as a "soft underbelly" for terrorist access to the United States grew especially intense as a result of the Ressam case. Although there is no evidence of a direct link between Canada and the terrorist attacks of 11 September 2001, the event has clearly highlighted the requirement for Canada to tighten its immigration, customs, and security laws.[43]

Thus America's pursuit of the RMA is raising more than directly related issues such as problems of allied interoperability. It is also having indirect implications, in that America's conventional superiority is prompting its adversaries to respond asymmetrically, particularly against the U.S. homeland. Canada's geographical tie to the United States is such that it will need to take measures to respond to the increased threat to North America. The state of Canada-U.S. foreign and defence relations over the coming decade will largely depend on the degree to which we support America's security agenda at home.

CHAPTER EIGHT

Canada and the RMA

The Department of National Defence (DND) and the Canadian forces have been aware for several years of the rapid advances in military and civilian technologies that are revolutionizing warfare. An in-depth examination of new technologies formed part of the process leading up to the *1994 Defence White Paper* and the white paper's stated objective of having armed forces that can fight "alongside the best, against the best" is an implicit reference to advanced capabilities. Nonetheless, the DND and the Canadian forces have only relatively recently begun to focus their attention on the RMA *in its entirety* and its implications for Canada. This chapter examines Canada's response to date, the extent to which the RMA is reflected in current Canadian defence policy, and measures Canada is already taking to respond to its technological, conceptual, and organizational elements. It then outlines a way ahead for Canada, looking first at the key contextual factors that are likely to frame any Canadian approach to the RMA and then examining their likely implications.

The chapter finds that Canada is taking preliminary steps to respond to the opportunities and challenges of the RMA but that many of these are as yet aspirations and stated future policy positions, rather than concrete achievements. Moreover, some technological and equipment trends of the Canadian forces would seem to go against the requirements of the future security environment. Selected and appropriate investments in the RMA are imperative and can do much to increase the capability of Canada's military forces. But they cannot obviate the inevitable trade-off between people and equipment. Ultimately, increased defence spending is necessary if the

Canadian forces are to make a concrete contribution to the achievement of Canada's future global security objectives.

POLICY STATEMENTS AND VISIONS

The revolution in military affairs first made it onto the agenda of the DND's highest-level departmental decision-making committee, the Defence Management Committee, in May 1998. Prompted in part by work that had been done earlier that year by the department's Directorate of Strategic Analysis, senior management directed that a newly created RMA working group develop a Canadian perspective on the RMA. The DND responded to this imperative with two high-level conferences. *Canadian Defence beyond 2010* (1998) focused on the impact of the RMA on various Canadian forces and departmental activities, including operations, science and technology, materiel, human resources, and the defence industry. A detailed *RMA Concept Paper* (1999) set out key conclusions and recommendations, one of which was the need to examine the tools for implementing the RMA. Accordingly, *Creating the Canadian forces of 2020* (2000) looked at concept development and experimentation and the integration of modeling and simulation across the DND and the Canadian forces, to determine how they can best be used to validate future force structure.

These concrete efforts to examine the RMA and methods of implementing it have been bolstered by the support of high-level departmental officials. The minister of National Defence has argued that "exploring and understanding the RMA is vital if Canada is to meet its future defence objectives."[1] The former chief of the defence staff has gone into more detail, arguing that Canada must focus its attention on elements of the RMA that help it implement its defence policy, such as conducting jointness and combined operations and ensuring interoperability.[2]

Canada's relatively recent focus on the RMA is reflected in the evolution of its official defence policy statements. The white paper of 1994 made no mention of the revolution in military affairs. This is perhaps not surprising, since even in the United States, which has been at the forefront of thinking about the changing nature of warfare, RMA technologies, doctrines, and concepts did not figure extensively in an official defence policy statement until *Joint Vision*

2010 was released in 1996. The term "revolution in military affairs" itself did not appear in a policy document until the *Quadrennial Defense Review* of 1997.

Beginning in 1997 some of the vision statements of the Canadian military services also began referring to the RMA. *Adjusting Course: A Naval Strategy for Canada* (1997) discussed many of the technological elements of the RMA, including improved sensors, more rapid transmission of information, stealth, and precision force. It also looked at the likely organizational effects of these technologies, such as the need for people with new skill sets and the requirement to redefine traditional command and control arrangements. *Canada's Army* (1998) briefly mentioned the RMA, some of its technological aspects, and the fact that the RMA is likely to influence how armies of the future are trained, organised, and equipped. It noted doctrinal trends towards increased joint and combined operations and the dispersion of forces on the battlefield, and it noted that technologies associated with the RMA will necessitate changes in military organizations and command and control structures.

Beyond the individual services, several DND-wide defence policy documents encompass objectives and milestones that resonate well with RMA-related technologies and concepts. *Shaping the Future of the Canadian forces: A Strategy for 2020* (1999, hereafter, *CF 2020*) highlights several key themes that will guide the future strategic direction of the Canadian forces. They include an increased focus on jointness, ensuring Canadian forces are interoperable with U.S. forces and capable of combined operations, and developing a force structure that is more rapidly deployable to crisis spots around the world – all of which figure as important elements of the RMA. *Defence Planning Guidance 2001* (2000) notes that rapid scientific and technological change is having a dramatic impact on weapons, equipment, and how military operations are conducted. Reflecting *CF 2020*'s vision, it outlines several objectives for change, including targeting leading-edge technologies and doctrines, enhancing deployability, and strengthening interoperability, that are consistent with the RMA. *Defence Plan 2001* (2001) refers more broadly to the ongoing challenge of the revolution in military affairs and lists five key change initiatives that are priority areas for the DND and the Canadian forces. The first two, "putting people first" and "optimising force structure to meet capability requirements," are central to the RMA. The following section examines in more detail the manner in which Canada is responding to the RMA's technological, conceptual, and organizational tenets.

RMA TECHNOLOGY

Precision Force and Precision-Guided Munitions

The Canadian forces have already invested in or have plans to acquire several RMA-related technologies. Most visible in recent operations have been the Canadian forces assets in the area of precision force and precision-guided munitions (PGM). In 1996 the air force began equipping several of its CF-18 *Hornet* fighter aircraft with the capability to conduct precision bombing, installing infrared sensors that allow pilots to see targets at night and laser designators that enable them to guide precision weapons to their target.[3] The air force also took delivery of its first precision munitions, which included GPU-10 and GPU-12 laser-guided bombs purchased from the United States. These measures meant that Canada was in a good position to take part in NATO's Operation Allied Force in and around Kosovo in the spring of 1999. Canadian fighters flew almost seven hundred combat sorties during the Kosovo campaign, dropping GPU-10 and GPU-12 precision-guided bombs, as well as Mk 82 "dumb" bombs.

The successful use of precision munitions in the Kosovo air campaign has prompted Canada to take steps to increase its PGM inventory. The air force plans to draw down its Mk 82 stocks while purchasing additional GPU-10 and GPU-12 munitions, as well as possibly the GBU-16 laser-guided bomb.[4] In addition, problems with the weather during the Kosovo campaign have prompted Canada to examine the acquisition of precision munitions that are guided by the Global Positioning System (GPS). While laser-guided systems rely on clear skies to allow a laser "lock" on a target, GPS-guided systems can "see" through the clouds, and they thus have an all-weather capability. The U.S. Air Force has already announced that it will not be procuring any more laser-guided weapons because of their weather limitations. Canada's air force has no immediate plans in place to acquire GPS-guided precision systems, but if it should do so in the future, one option may be the Joint Direct Attack

Munition, a relatively inexpensive GPS-guided system that proved very effective during the Kosovo and Afghanistan campaigns.

Battlespace Awareness and Control

Canada is also investing in advanced surveillance and command and control capabilities. Well aware of the growing and future importance of space for conducting effective military operations, the DND is acquiring its first dedicated military satellite communications capability. Currently, Canada relies on rental agreements with commercial satellites for long-range military communications. Under a memorandum of understanding with the United States, Canada is participating in the U.S. military's advanced extremely high frequency satellite system, which is due to have an initial operating capability in 2006, with four satellites up and running by 2008. Canada will develop its own dedicated military communications system to "piggy back" on the American satellites. Not only will the system provide secure data and voice communications for Canadian forces units world-wide, but by choosing to pursue a capability in an American system, Canada is ensuring future interoperability with the United States.

A second key program is the Joint Space Project, which primarily involves surveillance systems and sensors to monitor activity on the ground and in space. The project will give Canada a foothold in space and may also provide it with a means of contributing to America's proposed ballistic missile system.[5] *Defence Planning 2001* lists developing surveillance and communications technologies like the joint space project as part of one of its key change initiatives. The Canadian forces are also installing a Ground Moving Target Indicator on the *Radarsat II* (commercial) satellite, scheduled for launch in the next several years, which will be capable of tracking moving targets on the ground, such as missile launchers and armoured personnel carriers.

Meanwhile, the Canadian army is taking several steps to "digitize" the battlefield. The Iris Tactical Command, Control and Communications System is comprised of hand-held and vehicle-mounted radios operating in several frequency bands and will provide the army with a secure communications system that can be integrated with several subsystems. The Situational Awareness subsystem will permit vehicle and unit commanders to know where all friendly and enemy vehicles are with pinpoint accuracy day or night and in any weather. The battalion-and-above *Athene* battlefield management system automates such tasks as analysing terrain, processing and creating messages, and updating maps. And the omnibus ISTAR (intelligence, surveillance, target acquisition, and reconnaissance) project, launched in 1999, is being designed to integrate all these sensing, analysis, and transmitting technologies, potentially giving commanders the ability to "see" and "sense" all aspects of tomorrow's battlespace.

The Canadian navy has arguably been at the forefront of advanced command and control capabilities in the Canadian forces for several years. "The current Canadian vision of the Army describes battlespace awareness and c4I [command, control, communications, computers and intelligence processing], but the Canadian navy is already employing these aspects. In near and real time, we fuse information from a wide range of sources, developing a single coherent picture that is then shared amongst the fleet and with others."[6] Building on this, the navy has taken steps to improve its ability to control the battlespace, for instance by upgrading the satellite communications systems on its Iroquois-class *Trump* destroyers.[7] During Operation Allied Force this measure was cited by our allies as "the best example yet of a small navy adapting to a large task," because it gave the Canadian navy both "information on demand" (real-time information) and "a high degree of communications and interoperability fit" (interoperability) with our allies.[8]

A notable capability gap in the Canadian military's battlespace awareness and control capabilities lies in manned and unmanned aircraft. Whereas Canada's allies, including not only the United States but also Britain, France, Germany, and Australia, are developing and fielding unmanned aerial vehicles for surveillance and intelligence gathering, Canada has deferred the funds available for an Unmanned Aerial Surveillance Target Acquisition System.[9] Nor does Canada have any equivalent to the American Joint Surveillance Target Attack Radar System or the planned British Airborne Stand-Off Radar ground surveillance aircraft, although the Aurora incremental modernization project does include a Ground Moving Target Indication capability. Canada can be expected to participate in NATO's planned airborne ground surveillance system, just as

Canadian forces personnel are regular members of the crews aboard NATO's Airborne Warning and Control System aircraft, but such a capability is still several years away.

RMA DOCTRINAL AND ORGANIZATIONAL CHANGES

Jointness and Combinedness

Canada is also taking measures that are consistent with the RMA's key doctrines, some of which are bringing about important organizational changes. For example, Canada is striving to increase the "jointness" of its forces. Here, the DND is building on almost four decades of service integration. As a result of amendments to the National Defence Act in 1964, the service chiefs of staff were replaced with a single chief of defence staff, with executive authority over the three services. Additional changes in 1967 integrated the logistics, supply, and, to a certain extent, training systems, bringing a measure of rationalization and economy of forces. Unification the following year abolished the three services and created the Canadian forces. By taking these measures and, notably, by making a single deputy chief of defence staff responsible for the strategic direction of Canadian forces operations, Canada was well ahead of its allies in creating some of the key elements of a joint force. Indeed, in many ways, current allied efforts to promote joint doctrine are designed to match the support aspects of the Canadian forces' integrated and unified system.

Despite these steps, until recently "unification was, for various reasons, institutional and bureaucratic, but not operational."[10] In this context it is useful to make a distinction among three "levels" of jointness: joint-support and training, joint operations, and joint force planning. The steps taken in the late 1960s to a large extent achieved the goal of joint support and training. Actual operations, however, continued to be conducted along traditional service lines, essentially in isolation from one another. Maritime Command focused its efforts on protecting the sea lines of communication in the North Atlantic, Land Forces Command was primarily committed to the central front in Germany, and Air Command played a major role within NORAD. Even when it came to Canada's speciality during the Cold War, peacekeeping operations, the services tended to work separately, with contributions normally coming from one service. The nature of their primary tasks meant that the Canadian navy and air force worked extensively with their U.S. counterparts during the Cold War, while the Canadian army worked more closely with the German army. These patterns have meant that the navy and air force find themselves today somewhat more interoperable with the U.S. military than does the army and that Canada has gained far more experience in combined than in joint operations.

This is not to say that the services did not ever work together during the Cold War. The Maritime Air Group that participated in the navy's antisubmarine role was part of Air Command, as were the army's supporting helicopter and air transport elements. Today these air force assets are part of the First Canadian Air Division. These practical arrangements make apparent the potential of integration and unification to bring about jointness in operations: "In most other navies, the maritime air element is an integral element of that navy. In most other armies, the helicopter element of a brigade group is an integral element of that army. Such examples demonstrate that, in Canada, jointness already exists to a large degree."[11] As for joint force planning, although there is an organization within National Defence Headquarters that oversees force planning, much of it is still done by the individual services.

Over the past few years the Canadian forces have placed increased attention on promoting jointness. The 1994 Defence White Paper called for the forces to be able to deploy a joint task force headquarters as part of a multilateral operation anywhere in the world or in defence of a NATO member state. In response, the Canadian forces created a standing core of a deployable Joint Force Headquarters. In the event of a crisis, the headquarters, which is based on the headquarters of the First Canadian Division, would be augmented with appropriate cross-environmental staff.[12] In a combat situation the joint headquarters would exercise national command functions over the Canadian contingent and, depending on the situation, perhaps operational command over the forces of other countries, while in a peacekeeping situation the headquarters could, in principle, become the core of a multilateral operation. Nonetheless, Canada's attempt to lead a humanitarian relief operation in Africa's Great Lakes region in 1996, which was later abandoned due to changing circumstances, demonstrated the Canadian forces' still-limited ability to actually mount, deploy, and employ a joint task force. As a measure to address this shortfall, in 2000 the Canadian forces amalgamated their Joint Force Headquarters with the First Canadian Division. The DND's *2001–2002 Report on Plans and Priorities* places "transform[ing] the Joint Force Headquarters into a deployable C4I organization" high on its list of priorities for the modernization of force structure.

In terms of operational exercises, the services are focusing increasingly on jointness in their training exercises. In recent years the Maritime Coordinated Operational Training (MARCOT) exercises, held annually or biannually off the east or west coast of Canada, have incorporated elements of all three of Canada's services. In addition, *Defence Planning Guidance 2001* calls for the DND to develop a program of joint experimentation and concept development by mid-2002. The last decade has also seen some limited examples of joint Canadian forces operations. During the operation in Somalia, an Auxiliary Oiler Replenishment ship supported the Canadian contingent by providing sealift from Canada for military equipment and supplies and by acting as a floating headquarters.[13] More recently, both the air force and, to a lesser extent, the navy have worked with the army in operations at the lower end of the operational intensity spectrum. These have included *Op Assistance*, Canadian forces support to Manitoba during the Red River flood of 1997, and *Op Recuperation*, Canadian forces support for Ontario and Quebec during the ice storm of 1998. However, such involvement has usually been in a facilitating role, rather than as an actual joint operation.[14]

In responding to the RMA's doctrinal trends of increased joint and combined operations, Canada must bear in mind its particular needs and role in the world. The fact that the United States is moving towards a joint force structure may or may not be applicable to Canada, depending on the situation. For domestic operations it is important that the Canadian services be able to work together. But for overseas missions it is almost inconceivable that Canada would operate on its own. In these situations it will be far more important that the military services be able to work with their allied, and especially American, counterparts. It is instructive that while references to jointness appear as a component part of some of the change objectives of *Defence Planning Guidance 2001*, the document devotes an entire objective to strengthening military-to-military relations with Canada's allies and ensuring interoperable forces, doctrine, and C4I.

Interoperability

Central to the move towards increased jointness at the national level, and combined operations at the international level is the concept of interoperability. Within Canada the three elements of the Canadian forces are improving their ability to work together. They are, for example, adopting a command and control capacity that links the army, navy and air force information systems with the National Defence Operations Centre at National Defence headquarters. The Joint Command and Control Intelligence System uses the same software, terminals, protocols, and security features across the services and is expected to be in place by 2003.[15]

At the same time, all three of Canada's services are taking steps to enhance their ability to work with the militaries of their NATO allies, particularly the United States. Generally speaking, all investments in new technology are made with a view to ensuring interoperability in a multilateral context. The navy's satellite communications system upgrade program is meant to address interoperability problems created by the marked C4I gap between the U.S. navy and its allied counterparts, which was made apparent in NATO exercises. The Incremental Modernization Program for the CF-18 is aimed at improving the fighter's interoperability with other forces through, for example, installing new radios, friend or foe identification systems, and electronic warfare jamming equipment that is consistent with the equipment of Canada's allies. And the army's Iris Tactical Command, Control and Communications System will be fully interoperable with NATO command and control systems. More broadly speaking, the Canadian forces participate in NATO's longstanding standardization programs, which have as their goal multilateral standardization and interoperability of equipment, concepts, and doctrine, and they also participate in NATO's more recent Defence Capabilities Initiative, which focuses explicitly on improving interoperability among Allied forces.

Doctrinally, the navy already has a high level of interoperability with other NATO navies, especially the U.S. Navy, and this has made it particularly well suited for playing a supporting role, such as participating in sea control, within a multilateral coalition. The air force has worked closely with its U.S. counterpart through NORAD over many decades and has recently turned its attention to ensuring its forces are capable of integrating "as seamlessly as possible" into America's new Air Expeditionary Forces.[16] Canada's fighter aircraft were interoperable with those of the United States during NATO's 1999 Kosovo operation, although they lacked important technologies like a secure radio, Link 16, and GPS – all shortfalls that are being addressed in the upgrade program. Finally, to enhance interoperability with their U.S. counterpart, Canada's land forces have increased their participation in U.S. Army exercises.

More Rapidly Mobile and Flexible Ground Forces

In January 2000 the minister of National Defence announced the Canadian forces' intention to establish a rapid-reaction force available to either NATO or the United Nations to respond swiftly to global crises. The Canadian forces have already created a Disaster Assistance Response Team to respond to humanitarian crises and natural disasters and a Theatre Activation Team to deploy ahead of an overseas mission, to lay the groundwork for the main body of soldiers. The new rapid-reaction force would involve a battalion (1,000 to 1,200 troops) or one or two battle groups (1,250 to 1,500 soldiers each) and be capable of being in action within thirty days of receiving a request for contributions. Depending on the mission, the strike force could include *Coyote* reconnaissance vehicles, *Griffon* helicopters, LAV III armoured personnel carriers, *Hercules* transport aircraft, CF-18 fighter aircraft and the navy's frigates. The proposal is still in the examination stage, pending enhancements in strategic air- and sealift assets (discussed below). Solid improvements in Canada's rapid reaction capability are not expected before 2006.[17]

Defence Plan 2001 lists "significantly strengthening the [Canadian forces'] strategic mobility capability" as one of the DND's priority areas. One element of this is restructuring the army. Currently, Canada's land forces, with 19,000 people, organised into three mechanised brigade groups, are not well structured for quick deployment to areas around the world.[18] To address this problem, the army is being restructured into a force that is more readily deployable. A key benchmark, stated in both CF 2020 and *Defence Planning Guidance 2001*, is for vanguard and main contingency forces to be fully deployable to an offshore theatre of operations within twenty-one and ninety days respectively, utilising enhanced strategic air-and sealift capabilities. (Tellingly, this objective also appeared in the *1994 Defence White Paper* but remains yet to be achieved.) *Defence Plan 2001* directs the chief of the land staff to implement a transition plan to achieve a sustainable intermediate Army of Tomorrow structure by 2005. This is to involve a medium-weight, mechanised force and will lay the groundwork for a more fundamental army transformation, the Future Army, beyond 2011. Associated transition dates are to be articulated in *Defence Plan 2002*.

Increasing the rapid-deployment capability of Canada's ground forces requires lighter, more mobile military equipment that is still highly lethal and affords troops significant protection. To this end, the Canadian army is in the process of rationalising its fleet of armoured fighting vehicles and shifting the balance from tracked to wheeled vehicles. A large portion of the army's wheeled *Grizzly* Armoured Personnel Carriers (APC) and tracked M113A1 APCs are being phased out and replaced with the third variant of the wheeled Light Armoured Vehicle (LAV III), which is "faster, meaner and better armoured than its predecessors."[19] With its eight over-sized tires, the LAV III is more manoeuvrable and will provide increased operational mobility in theatre – attributes that are relevant to the dispersed, nonlinear nature of future military operations. It also comes fully loaded with an array of advanced sighting and laser technology.

Beyond this, the army has taken delivery of its "state-of-the-art" *Coyote* reconnaissance and surveillance vehicle. The *Coyote* can reach a top speed of 120 kilometres per hour and can see as far away as 14 kilometres, maintaining surveillance over a given area in all conditions of darkness and weather. The army is also planning to acquire a new armoured combat vehicle as a replacement for both the *Cougar* direct-fire support vehicle and the *Leopard I* main battle tank. One possible solution may be America's Future Combat System, expected to be available around 2012. Until then, however, the *Leopard I* is being upgraded with new thermal sights for improved night-fighting capability and computerised systems for enhanced weapons control. Defence analysts have criticised these expenditures, noting that the main battle tank – a heavy platform with limited battlefield manoeuvrability – is a Cold War relic ill-suited to the requirements of rapid deployment to a theatre and increased mobility within a theatre.[20]

Increased battlefield mobility also requires helicopters. The CH-146 *Griffon*, Canada's only land aviation helicopter, is used for everything from airlifting equipment and troops in theatre and conducting surveillance missions to search and rescue, medical evacuation, and assistance to civilian authorities. Its battlefield survivability and reconnaissance capabilities are set to be significantly improved with an advanced reconnaissance, surveillance, and target acquisition system. That said, the *Griffon* has its limitations. When it comes to lift, for example, it cannot transport the army's new lightweight field howitzer for more than 25 kilometres – short of the army's original requirement of 100 kilometres. Nor can the *Griffon* be considered a "combat" helicopter. Canada has never owned an attack helicopter like the U.S. *Apache*, whose chief function is to search and destroy enemy forces, the assumption always having been that such capabilities would be provided by allied units. The auditor general has concluded that the *Griffon* would be adequate "in mid-intensity conflict with the support of allies" – the implication being that it would be ill-suited to high-intensity conflict.[21]

A second element central to increasing ground force mobility is strategic air-and sealift. Currently the Canadian forces rely on thirty-two C-130 *Hercules* and five CC150 *Polaris* (Airbus A310) aircraft for airlift. Because of their limited numbers, however, these transports are often overworked, and Canada has on occasion had to "piggyback" on American transports or rent large Russian transport aircraft to deploy its troops and equipment into theatre. Equally significant is the fact that these aircraft are not large enough to transport some of the army's "outsized" vehicular cargo, such as the M109 self-propelled howitzer and *Leopard* tanks, both of which could be part of a Canadian contingency force. Capable of carrying only about seventeen tonnes at the relatively slow speed of 556 kilometres per hour, the *Hercules* is considered more of a tactical than a strategic aircraft. Beyond this, Canada is also limited in its sea-lift capabilities. It has no dedicated naval transport ships for moving equipment overseas, and the Canadian forces must rent commercial vessels for this purpose or transport their troops on allied vessels.

The DND's 2001–2002 *Report on Plans and Priorities* lists "develop[ing] options to enhance [Canadian forces'] deployability, including strategic lift," as the Canadian military's first priority for force-structure modernization. The air force is doing so in the context of its Future Strategic Airlift Project. It examined the options of purchasing or leasing a limited number of America's C-17s, purchasing some of the European A400M aircraft, or making arrangements with a commercial company to provide airlift, and recommended the first course of action. *Defence Plan 2001* directs the chief of the air staff to submit the project to the DND's Program Management Board for approval by July 2002. Meanwhile, the navy is addressing its sealift requirements, as well as the need to replace its ageing Auxiliary Oiler Replenishment ships, through its Afloat Logistics Sealift Capability project. The project is designing a ship that combines capabilities for fleet replenishment at sea, in-theatre support to joint forces ashore, and strategic lift for the army. Such vessels will significantly improve the Canadian forces' deployability and sustainability. *Defence Plan 2001* directs the chief of the maritime staff to submit the project to the Program Management Board for approval by July 2003, a target that may be "fast tracked" to allow construction to start as early as 2005.

Unmanned Combat

Canada has not yet made any concrete moves towards a third RMA doctrinal area, unmanned combat using unmanned combat aerial vehicles. To date Canada's attention has been focused on updating its manned fighters, the CF-18s, and looking at future options for new manned fighters. Since 1997 the DND has spent $15 million to make itself an "informed partner" with America's Joint Strike Fighter project. This allowed it to be privy to information on the aircraft's initial development, in the lead-up to the choice of a contractor. The Canadian military is now considering investing about $750 million over the next decade to participate in the next phase of the project. This would give Canada options to purchase the aircraft and would allow Canadian firms to participate in the industrial development and production of the aircraft.[22]

Littoral Warfare

Responding to the requirements of the new international security environment, Canada's navy has begun changing its focus from being an open-ocean, "blue-water" force to becoming one that can also conduct littoral operations in shallower, potentially more dangerous waters in support of joint forces ashore. The Afloat Logistics Sealift Capability, for example, reflects this change in focus and will enable the navy to better support Canadian forces units ashore.

Nonetheless, the navy has none of the major tools for sea-power projection onto land. For the U.S. Navy, such capabilities include aircraft from carriers, ship-to-shore cruise missiles, and the amphibious landing capabilities of the marines. Considering the utility of land-attack cruise missiles in recent peace enforcement missions in Bosnia (1995) and Kosovo (1999) and in the war against terrorism (2001), it may be prudent for Canada to consider investing in such weapons. While America's *Tomahawk* cruise missile is probably out of the question due to cost, a feasible option may be the Standoff Land Attack Missile. The navy's forthcoming strategy document, *Leadmark: The Navy's Strategy for 2020*, argues that the navy must enhance its ability to strike land targets ashore.[23] Such precision fire support would not be intended to create an amphibious assault capability but rather to strike at ground targets for the purpose of protecting Canadian troops and ships near shorelines.

THE WAY AHEAD

It is clear from the foregoing that Canada has already taken some initial steps to respond to the opportunities and challenges presented by the revolution in military affairs. In terms of technology, the Canadian forces is focusing on precision-guided munitions, particularly for the air force, and advanced battlespace awareness and command and control capabilities. Doctrinal and organizational changes to create a more rapidly mobile ground force and to enhance the Canadian forces' ability to conduct joint and combined operations are also in their preliminary stages. These changes are being supported by measures to acquire lighter, more mobile equipment and by incorporating advanced technologies that promote interoperability into units and equipment. It is also significant that the DND has placed front and centre in its most recent planning documents the priority of strengthening the Canadian forces' ability to recruit, train, and retain highly skilled personnel. These are the people that will be at the centre of a future Canadian RMA force.

As is the case with its NATO allies in Europe, however, many of Canada's efforts to respond to the RMA are as yet a stated policy for future direction, rather than a concrete achievement. A rapidly deployable ground force awaits the acquisition of strategic air-and sealift capabilities, still many years away. It will also be some time yet before Canada has a dedicated military satellite capability. And despite its centrality to littoral warfare, a land-strike capability for the navy is a distant aspiration. Moreover, some of Canada's technology and equipment trends go against likely future security requirements. The Canadian forces have stocks of laser-guided munitions but no immediate plans to purchase the satellite-guided weapons that proved central to recent operations. Nor is it taking advantage of the opportunities presented by unmanned aerial vehicles to as significant degree as are many, indeed all, of its major allies. And the Canadian forces is upgrading its heavy tanks and considering a new, stealthy manned fighter aircraft, even though these platforms are unlikely to be well suited to the future security environment.

The Role of the Canadian Forces

In continuing to develop Canada's approach to the RMA, policymakers may want to bear in mind two important contextual factors. The first of these must inevitably be the answer to the question, What do we want our forces to do? The 1994 *Defence White Paper* spells out three roles for the Canadian forces: the protection of Canada, the defence of North America in cooperation with the United States, and contributing to international security. The first two roles can be taken as a given: they have figured, in some form or another, in defence white papers dating back to at least 1964, and it is inconceivable that they would not be fundamental components of future defence policy – particularly in light of the terrorist attacks on the United States in September 2001 and of the increased saliency of homeland defence. The notion of "contributing to international security," too, is an enduring feature of Canadian defence policy, but here there has always been room for debate as to the exact meaning and content of the phrase. Views have ranged in the past from focusing on overseas development assistance to fielding a lightly armoured military force that can participate in low-risk UN peacekeeping and observer missions, to maintaining high-readiness, general-purpose, combat-capable forces that can take part in high-intensity warfare missions.[24]

The 1994 *Defence White Paper* incorporated both these latter missions, calling for a multipurpose, combat-capable force that can respond to a range of operations, including not only peacekeeping and observer missions but also operations enforcing the will of the international community and collective defence. "It is only through the maintenance of such forces," the white paper argues, "that Canada will be able to retain the necessary flexibility and freedom of action when it comes to the defence of its interests and the projection of its values abroad."[25]

Despite these words, there is a perception within the U.S. defence community that Canada has focused its defence policy almost entirely on peacekeeping.[26] Canada's leading role in this area throughout the Cold War is no doubt a source of this perception, as may be its more recent foreign policy focus on "soft power" and "human security." Regardless of the origins of the perception, in responding to the RMA Canada will want to keep its focus on the white paper mandate, identifying and investing in those technologies, doctrines, and organizational concepts that are applicable to both low-and high-intensity operations. "The solution to the dilemma for the [Canadian forces] will be in developing a force structure that reflects suitable capability for the most likely forms of employment [i.e., peace support operations], while retaining the ability to integrate in a meaningful way with the RMA forces of the United States in confronting the most dangerous contingencies."[27]

Canadian Military Capabilities

Notwithstanding the white paper mandate, there is some question about Canadian military capabilities. In the decade following the end of the Cold War the size of the Canadian forces fell by a third, from 87,000 to just under 60,000 personnel. Canadian defence spending also fell significantly, to the point that in fiscal year 1999–2000 Canada contributed just 1.1 percent of its gross domestic product to defence, compared to the NATO average of 2.1 percent. Reductions in force size and defence spending, in turn, appear to have had a direct impact on Canada's ability to maintain its military capability. A report of the auditor general in 1998 stated that repeated budget cuts had had a serious impact on both the numbers of people in uniform and the state of capital equipment.[28] In 1999 an academic study of the Canadian forces argued that the army could no longer meet its white paper goals and that the navy and air force were also at risk of losing the ability to meet their objectives.[29] Although military leaders have argued that the forces are more combat capable than they were a decade ago during the Gulf War, this point remains disputed by other defence experts.[30]

Implications

Thus, Canada enters the RMA debate against the backdrop of a stated intention of maintaining forces for the range of military operations and the unanswered question of whether reduced defence spending and force reductions since the end of the Cold War have significantly diminished its military capabilities. At the political level this could cause Canada to be increasingly marginalized in international forums. "Because we cannot expect our political influence in global and regional security arrangements to be significantly out of proportion to our military contributions," the *1994 Defence White Paper* argues, "we must make the required investment in our armed forces."[31] Already there is a sense among analysts that Canada's reduced military expenditures and, potentially, capabilities, have resulted in declining international influence.[32]

Investments in defence are also needed to avoid operational marginalization. "[I]f Canada wants operational influence within a coalition/alliance," argue some RMA analysts, "its forces must be capable of participating in a salient way."[33] In part this means ensuring its forces remain interoperable with those of its closest and most important ally. But it also involves maintaining a sufficient number of forces to form a combat-relevant force.

Opportunities for the Future

Against the backdrop of these contextual factors and implications, the revolution in military affairs offers Canada a number of opportunities. Broadly speaking, the nature of the RMA is such that it has the potential to provide relatively greater benefits to small to medium powers, such as Canada, than to large powers like the United States. This is because the RMA is changing the fundamental components of power. Previously, the most important elements of a country's military potential were its population and the strength of its economy. Increasingly, however, it is the quality of its standing forces, as opposed to the population as a whole, and of its high-tech sector, as opposed to the economy as a whole, that will be the determining factors. In this sense, Canadian policymakers should view the RMA as much as an *opportunity* as a *challenge*. Whereas the large mass army of the industrial era was out of reach for Canada, the smaller, more technologically capable force of the information age is within the realm of possibility.

In more specific terms, the RMA offers Canada opportunities to address the concerns noted above. In an era of restricted defence budgets, Canada is committed to maintaining the capability to respond to high-and low-intensity tasks. Selected investments in the RMA can enhance our ability to do both. Advanced C4I and intelligence, surveillance, and reconnaissance systems, for example, are highly relevant to the efficient conduct of operations at all points along the spectrum of conflict. Unmanned aerial vehicles for reconnaissance have proved their worth in a range of conflict scenarios, whether it be the high-intensity battlefield of the Gulf War, the medium-intensity NATO operation in and around Kosovo, or the low-intensity task of monitoring implementation of the Dayton Peace Accords. Strategic lift is another good example of a capability that is suited to both high-and low-intensity tasks, while precision-guided munitions are clearly relevant both to warfighting scenarios and to the peace-enforcement aspects of peace-support operations. All of these elements are also central to conducting the war against terrorism, as is the development of highly lethal yet rapidly deployable ground forces that are very mobile on the battlefield.

It follows that selected and appropriate investments in the RMA can go a long way towards increasing the capability of Canada's military forces. Indeed, "the RMA might well be the only way to get increased efficiency from the smaller forces planned since the end of the Cold War."[34] Such investments are imperative if the Canadian forces are to remain interoperable with their U.S. counterparts. But the reallocation of funds away from Cold War systems and towards those that are central to the RMA can only do so much. Ultimately there must be a trade-off between numbers and quality of troops and quantities and sophistication of platforms, since the same pot supplies money for both people and acquisitions. To ensure this trade-off is set above the line of operational and political marginalization, there is no alternative to increased defence spending, however modest, a fact that was implicitly recognised in the federal government's budget of February 2000 and is especially acknowledged in the wake of the terrorist attacks on New York and Washington. Such continued action is necessary if the Canadian forces are to make a concrete contribution to the achievement of Canada's future global security objectives.

Conclusion

A revolution in military affairs is under way that holds the potential to dramatically change the character of warfare over the next two to three decades. The information revolution in the civilian world is driving rapid advances in military technologies, most notably in computers, telecommunications, sensors, and precision-guided munitions. Although such advances have been under way since the 1970s, in the wake of the Cold War they have combined with factors that are giving ongoing evolutionary change a revolutionary "take-off" capacity. Cuts in defence budgets have brought about significant reductions in the size of military forces in Western countries, prompting a need to emphasise qualitative force attributes. The new international strategic environment is characterised by a range of threats and risks, resulting in a requirement for forces to be able to respond rapidly to a wide spectrum of regional conflict scenarios. At the same time there is a lower tolerance among Western publics and governments for military and civilian casualties. This combination of factors has given military establishments impetus to translate new military technologies into doctrinal and organizational change, thereby setting the stage for a true RMA.

The United States is at the forefront of revolutionary military developments. In the 1970s and 1980s it fuelled the RMA's technological advances with its "offset" strategy of using superior military technology to balance the Soviet Union's quantitative advantage in military forces and equipment. In the 1990s it began to incorporate advanced military technologies into new military doctrines and organizations. An assessment of current U.S. force transformation efforts, which have as their explicit objective the creation of an "RMA force," indicates that the next few years is unlikely to see a dramatic leap into high-tech warfare. Nonetheless, the increased political will for change both before and after the terrorist attacks of 2001 is such that the ongoing evolutionary force-transformation process has real potential to culminate in revolutionary change over the next two to three decades.

Canada's allies are taking several steps to respond to the RMA. Technologically, there is a strong focus on advanced surveillance, command and control, and precision-strike capabilities. Organizationally, the trend is towards professional, highly trained personnel and the creation of smaller, more modular units that facilitate both joint and combined operations. Conceptually, an important overarching doctrinal theme is the creation of "expeditionary" forces, that is, forces that are lighter, more rapidly mobile, and deployable, yet still highly lethal. The Canadian military is placing its emphasis on similar areas in its policy positions and acquisitions projections.

But an examination of current capabilities reveals that many of these measures – in Canada and Europe – remain as yet aspirations or stated policy for future direction, rather than concrete achievements. European members of NATO are not incorporating advanced technologies into their military systems quickly enough to stem an evident and growing technology gap between the U.S. military and its European counterparts. Canada is in a similar position, ameliorated only by the somewhat closer Canada-U.S. bilateral defence relationship since World War II. The gap, in turn, is calling into question the ability of the European and Canadian militaries to function with their American counterparts in the future. Although problems of compatibly have been an issue for the alliance since its inception, the difference today is that the U.S. military's technological advancements hold the possibility of completely eclipsing those of its allies. Moreover, American officials have expressed concerns that in the overall spectrum of conflict European allies may be concentrating on the lowest 50 percent of missions, while leaving the other (high-intensity) 50 percent to the United States. Likewise, there is a perception that Canada's military is focusing primarily on peacekeeping tasks.

To some extent, however, this distinction may be an artificial one. Certain aspects of the RMA have only limited application to peace support operations or are useful only at the very high end of the range of missions. There are also units central to peacekeeping – and to combating international terrorism – that will be in insufficient supply if the focus is solely on force attributes for high-intensity conventional warfighting. Examples include construction engineers and special operations forces respectively. But, generally speaking, many aspects of the RMA are applicable to both peace support operations and the war against terrorism. Relevant technologies and doctrines range from precision-guided munitions and advanced surveillance and command and control technologies and platforms to joint forces, littoral warfare, and armies that are more rapidly deployable, yet highly lethal and mobile on the battlefield. The best way of bridging the technology and capability gap between the United States and its NATO allies is for Canada and Europe to focus on those elements of the RMA that are suited to both peace-support missions and high-intensity war, whether this latter category involves a conventional or a terrorist foe.

Either by design or intent, this is exactly what Canada and key European members of NATO are doing in their stated defence policy positions for the future. It follows, then, that when it comes to allied responses to the RMA, the shortfalls lie less in the area of policy than in the area of implementation. Mechanisms have been put in place to try to boost implementation, most notably NATO's Defence Capabilities Initiative. However, progress has been limited. Part of the explanation can be found in the manner in which defence funds are put to use. In some countries Cold War defence infrastructures continue to take up funds that could better be put towards acquisitions or research and development. Moreover, scarce resources continue to go to upgrading and purchasing platforms that military trends indicate are unlikely to be well suited to the future security environment. Although the United States faces a similar set of issues, Canada and Europe do not have the same kind of leeway in their defence budgets that America has to still be able to devote substantial funds to a future force.

Reprioritizing within defence budgets can only do so much. Increasing the portion of a nation's defence spending that is devoted to equipment acquisitions is a step in the right direction, but ultimately this requires an important trade-off between numbers and quality of troops and quantities and sophistication of platforms –

since the same pot supplies money for both people and acquisitions. For Canada and European members of NATO, troops *and* equipment need to be set above a certain level if these countries are to make a meaningful contribution to future coalition operations. In the end, if Canada is to avoid political and operational marginalization in international forums and multilateral missions and if Europe is to enjoy true partnership and shared decision-making power with the United States in military operations, there is no alternative to increased defence spending, however modest. Only then will America's NATO allies, including Canada, be able to meet the challenges and exploit the opportunities presented by the revolution in military affairs.

Epilogue

Four months after the tragic events of 11 September 2001 the parameters of a new international order and the sorts of military forces that will best be able to operate within this order are starting to take shape. A decade ago the end of the bipolar competition between the United States and the Soviet Union left practitioners and thinkers working in the area of political and military affairs grappling with the nature of the new international security environment. Defying a title of its own, the "post–Cold War era" seemed to offer no global paradigm, or organizing principle, around which the United States and its Western allies could centre their foreign and defence policies. Had we reached the "End of History," of humanity's ideological evolution towards Western liberal democracy, and hence could we safely prepare and organize our military forces for conflicts between states "still in history" and between those states and the states at the end of history?[1] Or did the future portend the primary source of conflict to be cultural, and therefore should we look to different civilizations as the relevant framework for analysis and to their boundaries as the most likely scene of conflict?[2] Or was it really the environment that would be "the national security issue of the twenty-first century," with the political and military impact of surging populations, spreading disease, scarcity of resources, refugee migrations, water depletion, and air pollution revealing the order of things to come?[3] Although each thesis offered a compelling argument and some explanatory capacity, none emerged as an overall systematic structure that could satisfactorily guide policymakers.

The terrorist attacks on the United States have changed all this, bringing a measure of clarity to the international security environment. The attacks have solidified concepts of threat that had been gaining ground for half a decade. They have catalyzed trends in military technologies, doctrines, and organizations that can trace their origins to the 1970s. More than anything, they have provided the United States and its NATO allies with a new means of ordering foreign and defence policy priorities.

CONCEPTS OF THREAT

America's *Joint Vision 2010* of 1996 was the first official U.S. defence policy document to refer to possible asymmetrical counters to conventional American military superiority that could target American vulnerabilities. The *Quadrennial Defense Review* of 1997 elaborated, arguing that adversaries could use asymmetric means – including ballistic missiles, weapons of mass destruction, terrorism, and information warfare – to attack Americans not only abroad but also at home. Outside experts, including the National Defense Panel of 1997 and the U.S. Commission on National Security/Twenty-first Century of 1999–2001, saw the U.S. homeland as being particularly vulnerable and recommended several measures to deal with the new and growing threat.

Ultimately the response to the asymmetric threat will involve both a domestic, or "homeland," and an international security element. Homeland security, in turn, has both a homeland defence and a civil-support aspect, with the former involving warfighting missions where the military is clearly in the lead and the latter involving civilian-led operations that may need military support in a coordinated federal response. For homeland defence the U.S. defence secretary has already given various U.S. military commanders additional homeland-related authorities. For increased U.S. military support to civil authorities the effect on force structure and organization is still being determined but will almost certainly involve an increased role for the National Guard and other reserve forces. A primary area of emphasis for America's new Office of Homeland Security – and for any Canadian counterpart – will be simply to delineate the military's role in homeland security.

CATALYZING TRENDS

It is the international security element of the response to terrorism that is most directly related to the revolution in military affairs (RMA) and that is catalyzing trends in military technologies, doctrines, and organizations. In the war on terrorism in Afghanistan almost no area of the RMA was left untouched. The campaign demonstrated the utility of advanced precision munitions and the doctrines of precision force and disengaged combat. Some 60 percent of the munitions dropped on Afghanistan were precision-guided, compared with about 6 percent in the Gulf War and 35 percent during the Kosovo air campaign.[4] Their accuracy enabled the United States military to do much of the work of combating international terrorism from safe standoff distances.

Advanced surveillance, reconnaissance, and command and control capabilities were also central, with Unmanned Aerial Vehicles (UAVs) playing a particularly crucial role. Geared towards monitoring troop movements, the *Predator* adapted easily to the task of tracking terrorists and was used both for determining air strike targets and guiding and protecting U.S. ground forces. The *Global Hawk* strategic UAV, with its cloud-penetrating sensors, was also used extensively for surveillance and reconnaissance.

The campaign reinforced the naval trend towards littoral warfare and projecting power from the sea onto land. U.S. Navy surface ships and submarines launched *Tomahawk* cruise missiles against Afghan targets, while carrier-based fighter aircraft dropped precision-guided bombs. Meanwhile, B-1 and B-52 strategic bombers based on the island of Diego Garcia in the Indian Ocean, as well as stealthy B-2 strategic bombers based in Missouri, delivered "smart" weapons such as satellite-guided Joint Direct Attack Munitions and Standoff Land Attack Missiles, strengthening the concept of long-range force projection and stealth.

The war put into practice for the first time the revolutionary doctrine of unmanned combat using Unmanned Combat Aerial Vehicles (UCAVs). *Predator* drones armed with precision-guided missiles carried out dozens of strikes in Afghanistan with a reportedly impressive rate of accuracy, prompting experts to predict that some 90 percent of combat aircraft will be unmanned by 2025.[5]

The advent of UCAVs for precision strike and the use, as in Kosovo, of long-range strategic aircraft for this mission reinforced the doctrinal trend away from short-range tactical aircraft. U.S. Air Force fighters found themselves all but shut out of the war against terrorism because of the lack of basing rights in Central Asia. And although carrier-based fighters proved integral to this first war of the twenty-first century, they may be less so in a future conflict against a foe that could threaten carriers with antiship missiles or refueling aircraft with antiaircraft fire.

All wars are ultimately won on the ground, and the campaign in Afghanistan was no different. Highly mobile ground forces that pack enough firepower and electronic support systems to operate independently are central to RMA doctrine, and they were fundamental to success in Afghanistan. Strategic lift provided rapid deployment, while helicopters helped ensure tactical mobility. As was the case in Kosovo, local ground forces – in this case the Northern Alliance – were also a decisive factor, but their work would not have been possible without air power first striking entrenched Taliban positions. At the same time, the use of special operations forces to locate targets and call down precision firepower from the air force and navy confirmed the RMA's doctrinal trend of increased "jointness" among military services.

Was the war on terrorism in Afghanistan a "one-off" or a preview of things to come? Will investing in the technologies and forces that were useful in this conflict – described as a major theatre war by Pentagon officials – be akin to preparing for the last war, the likes of which we are unlikely to see again?

It is true that there is unlikely to be a war that is similar to the war in Afghanistan again. As U.S. secretary of defense Donald Rumsfeld said in a news conference in December 2001, the conflict was distinctive because of the geography of the country and the particular circumstances involved. But this was true too of the Gulf War and of the peace enforcement operation in Kosovo. What is striking about these conflicts is the number of revolutionary technologies and doctrines that were applicable to three such very different scenarios. Many changes associated with the RMA are relevant across the spectrum of conflict and thus are well suited to responding to an international security environment characterized by diverse and unpredictable threats and risks.

FORCE TRANSFORMATION AND THE RMA

Not surprisingly, the Pentagon is now promoting several key elements of the RMA. Efforts have been launched to boost funding and accelerate delivery schedules for both tactical and strategic UAVs. The U.S. Navy hopes to restart its *Tomahawk* cruise missile production line and add hundreds more of these missiles to its inventory. A debate is underway in the Pentagon about the future of air force tactical fighters. And the U.S. Army is seeking to hasten deployment of its new light brigades and is looking at options to speed development of its future combat system. "This war is going to give you the revolution in military affairs," argued one well-known expert in military innovation in the opening stages of the war against terrorism.[6]

This does not mean that U.S. military forces will be transformed overnight. Less than two months after the terrorist attacks, the Department of Defense awarded its biggest contract ever – the Joint Strike Fighter contract – to build three versions of a manned short-range tactical fighter. A bolder, more logical step would have been to cancel the land-based version. Moreover, other legacy systems, such as the army's *Crusader* self-propelled howitzer, are now being touted as important for an antiterrorism campaign, even though they are very difficult to transport quickly.

Nonetheless, the new international situation has boosted transformation by a marked degree in both political and budgetary terms. Before the attacks of 11 September Secretary Rumsfeld faced an uphill battle to transform the U.S. military. In their aftermath his position has been vindicated and transformation has come to be viewed as imperative. President Bush voiced this urgency in December 2001 in a speech to cadets at the Citadel military academy, stating explicitly that "a revolution in [the U.S.] military" is needed to attack and defeat terrorism. The additional funds provided to the Department of Defense following the terrorist attacks will increase short-term opportunities for developing and deploying systems associated with the RMA. Over the long-term, procurement spending can be expected to shift away from traditional ships and tanks and fighter aircraft and towards tools and technologies central to the war against terrorism and, by extension, to the RMA. In this way transformation will proceed at an accelerated pace even as traditional programs persist. The ultimate result will be a revolution in military affairs.

IMPLICATIONS FOR CANADA AND NATO

For America's allies in NATO, including Canada, the effect will be to place in even sharper light issues and concerns that were already preoccupying the alliance before the terrorist attacks. Tellingly, the United States made almost no military requests of NATO during the war on terrorism in Afghanistan, choosing instead to prosecute the war essentially on its own, with assistance from Britain and to a smaller degree from other allies (including Canada). With the exception of the Airborne Warning and Control System aircraft sent to patrol North American skies, NATO's contribution was confined to providing political support. This is not an insignificant role – after all, a key aspect of NATO has always been the political dimension in which allies can engage in consultation and dialogue about common threats and risks. Nonetheless, the conflict demonstrated how far NATO has travelled from being able to operate as a military alliance in collective defence of an ally.

Increased U.S. defence spending and stepped-up transformation efforts in the wake of the terrorist attacks will reinforce this trend by widening further the technology and capability gap between the United States and its North Atlantic allies. Unless Canada and European members of NATO make sustained and significant investments in their armed forces, they can expect to find themselves unable to contribute to a future high-intensity operation, whether it be a peace enforcement mission or a major theater war – which are distinguishable only by objectives and not by the tools involved. Some commentators may argue that this is a desirable position. What's wrong with staying home and focusing on domestic concerns and leaving international security problems to the United States? is a rhetorical question sometimes heard in academic circles. The answer, of course, is that there is nothing wrong with this position if that is the worldview one holds. But for those policymakers and members of the general public who want their country to be able to make a tangible contribution to international peace and security and to promote global security objectives, such a position is unacceptable.

This is doubly the case when one considers that the terrorist attacks on the United States have demonstrated vividly an eroding distinction between the national and international dimensions of security. To leave the problem of international terrorism to fester in a far-off country is to increase the probability that it will eventually land on your doorstep. In this context, the downside to not investing in some of the key RMA technologies and doctrinal changes that have been discussed in this book and have proven relevant to the war against terrorism goes well beyond Canada's political and military marginalization and Europe's lack of shared decision-making power with the United States in military operations. It involves not addressing an issue that poses a real threat to members of the North Atlantic alliance, both collectively and individually.

The Cold War was a political, military, and ideological struggle against a common foe that provided an overall framework for ordering foreign and defence policy. Militarily, NATO members recognized they could not opt out of responding to the threat and took the necessary measures to do so. The challenge today, in this "post–September 11 era," is to recognize the same in the concrete terms of increased military capabilities. A new order has emerged, one that represents no less an all-encompassing struggle, one that is characterized by new forms of threat, and one that demands new means of responding.

Ottawa, January 2002

Notes

CHAPTER ONE

1 Rogers, "Military Revolutions," 22.
2 Hundley, *Past Revolutions*, 9. For example, if air-and/or sea-launched precision munitions could be used to unleash a massive barrage of fire power
  against army formations, thereby stopping the advance in its tracks, this
  would effectively eliminate heavy armour as a viable means of waging war
  in the twenty-first century.
3 Lambeth, "Technology Revolution," 75.
4 Matthews, "Improved Space Assets," 44.
5 Willis, "Field Artillery," 12.
6 Matthews, "Improved Space Assets," 44.
7 "The Unmanned Inevitability?" 14.
8 Lambeth, "Technology Revolution," 69.
9 Evers, "USN's Next Generation Destroyer," 8.
10 Foxwell and Lok, "Approaching Vanishing Point," 43.
11 Roos, "Disappearing Act," 62.
12 Cooper, "U.S. Stealth Enhancements," 1.
13 Friedman and Friedman, *The Future of War*, 176.
14 Watkins, "USAF System," 12.
15 Clark, "U.S. Army," 6.
16 "USA Takes Latest C2 System into the Field," 38.
17 Lambeth, "Technology Revolution," 72.
18 Davis et al., *The Submarine*, 31.
19 Holzer, "U.S. Navy," 8.
20 Macgregor, *Breaking the Phalanx*, 60.
21 "The Future of Warfare," *Economist*, 8 March 1997, 22.
22 Nye and Owens, "America's Information Edge," 27.
23 Mazarr et al., *Military Technical Revolution*, 35.

24 Seffers, "Land Forces," 14.
25 Seffers, "US Army Study," 8.
26 Mazarr et al., Military Technical Revolution, 34.
27 Lambeth, "Technology Revolution," 66.
28 Holzer, "Questions," 4.
29 Hewish, "Coming Soon," 30.
30 "Robots Herald New Era," 22.
31 Hewish, "Bird's-Eye View," 55.
32 Hewish, "Coming Soon," 31–2.
33 Andrew Krepinevich, executive director of the Center for Strategic and Budgetary Assessments, as quoted in Walsh, "USAF," 26.
34 Cohen, "A Revolution in Warfare," 46.
35 Ibid., 47.
36 "Keep Pace with Technology," 18.
37 Macgregor, Breaking the Phalanx, 88.
38 Holzer, "U.S." 34.

CHAPTER TWO

1 Professor Robert L. Pfaltzgraff Jr, in lectures to students at the Fletcher School of Law and Diplomacy, 1994–95.
2 Toffler and Toffler, War and Anti-War.
3 Krepinevich is executive director of the Center for Strategic and Budgetary Assessments, Washington, DC.
4 Krepinevich, "Cavalry to Computer".
5 Murray is professor emeritus at Ohio State University.
6 Murray, "Thinking about Revolutions".
7 Clifford Rogers is assistant professor of history at the United States Military Academy, West Point.
8 Rogers, "'Military Revolutions.'"
9 Krepinevich, "Cavalry to Computer," 30–1.
10 Murray, "Thinking about Revolutions," 73.
11 Hundley, Past Revolutions, 13.
12 Krepinevich, "Cavalry to Computer," 37.
13 Perry, "Desert Storm and Deterrence," 68–9.
14 Blaker, Understanding the Revolution, 5.
15 Davis, "An Information-Based Revolution," 46.
16 Blank, "Preparing for the Next War," 20–1.
17 Blaker, Understanding the Revolution, 5–6.
18 Perry, "Desert Storm and Deterrence," 77, 80–1.
19 Blaker, Understanding the Revolution, 7.
20 Ibid., 8.
21 Mastanduno, "Preserving the Unipolar Moment," 51.
22 Ibid., 66.

23 Tucker, "The Future of a Contradiction," 20.

24 Gouré, "Military-Technical Revolution?" 175.

25 Luttwak, "A Post-Heroic Military Policy."

26 O'Hanlon, *Technology Change*, 46. O'Hanlon discusses several other technological limits to the RMA, including bandwidth limitations to advanced communications and internal combustion engine efficiency limitations to increasing the speed of platforms.

27 Thomas, "Kosovo."

28 Blank, "The Illusion of a Short War," 138.

29 Owens, "Technology," 68.

30 Biddle, "Victory Misunderstood."

31 Blank, "Illusion of a Short War," 140.

32 Henley, "The RMA after Next."

33 Ibid.

34 Dr Michael Margolian, Department of National Defence, Ottawa, Canada, 12 April 2001.

35 O'Hanlon, *Technology Change*, 29.

CHAPTER THREE

1 O'Hanlon, "High Technology?" 76–8.

2 Henley, "The RMA after Next."

3 Rogers, "'Military Revolutions,'" 31.

4 Andrew Marshall, director of the Office of Net Assessment, Office of the Secretary of Defense, statement before the Senate Armed Services Committee, 5 May 1995.

5 Owens, "Technology," 67.

6 Flournoy, *Report*, 14.

7 O'Hanlon, *Technology Change*, 27.

8 *Science, Technology and Military Strength*.

9 The others the Pentagon is pursuing are "A Focused Modernization Effort," "Exploiting the Revolution in Business Affairs," and "Hedging against Future Threats."

10 The first two pillars are "Shaping the International Environment" and "Responding to the Full Spectrum of Crises."

11 See sections 3 and 7 of the QDR, "Defense Strategy" and "Transforming U.S. Forces for the Future," respectively.

12 See sections 1, 2, and 5 of the Quadrennial Defense Review Report, released 30 September 2001.

13 Defined as "the capability to collect, process, and disseminate an uninterrupted flow of information while exploiting or denying an adversary's ability to do the same."

14 Cohen, *Annual Report*, 178.

15 Holzer, "Experts," 12.

16 See, for example, *Transforming Defense,* 49.

17 Cohen, "Defending America," 47.

18 Krepinevich, *Wither the Army.*

19 "The New Air Force Line-Up."

20 Wirtz, "QDR 2001," 54.

21 QDR, section 7, "Transforming U.S. Forces for the Future."

22 Goodman, "Aerospace Force," 48.

23 McGinn, "Lessons from the National Defense Panel," 21.

24 *Why No Transformation?*

25 Bender, "DOD," 7.

26 Seigle, "USA Forms Joint Force Command," 4.

27 Donnelly, "Revolution?" 23.

28 Hundley, *Past Revolutions,* xxii.

29 Donnelly, "Revolution?" 25.

30 Correll, "Aerospace Power," 2.

31 Lewis and Roll, "Quadrennial Defense Review 2001," 73.

32 Holzer, "Think Tank Defense Vision," 40.

33 Donnelly, "Revolution?" 24.

34 Ibid., 26.

35 "U.S. Military Spending."

36 Bender, "Budget Pressures," 8.

37 Gouré, "The Resource Gap," 39.

38 United States Commission on National Security/Twenty-first Century, *Road Map,* xii.

39 Cohen, "Defending America," 44.

40 Snodgrass, "The QDR."

41 Cohen, "Defending America," 44.

42 Bush, "A Period of Consequences."

43 "Remarks by the President."

44 Allen, "Bush Vows Spending."

45 "New Arms for a New World," 31.

46 Ricks, "For Rumsfeld, Many Roadblocks."

47 Scarborough, "Pentagon to Focus on Defense of U.S. Soil."

48 O'Hanlon, "Prudent or Paranoid?"

49 Cohen, "Defending America," 45.

50 Gertz, "Rumsfeld Recalls Cold War."

51 Canahuate, "U.S. Army Considers Acceleration of FCS Program."

52 Canahuate, "Pentagon to Accelerate Development."

CHAPTER FOUR

1 Cook, "UK to Accelerate Watchkeeper UAV Program."

2 Barrie, "Britain May Speed Up UAV Project," 44.

3 Thomas, *Military Challenges,* 15.

4 Adams, "British Set Example," 15.

5 United Kingdom defence secretary George Robertson, Intervention at NATO Defence Ministerial, Vilamoura, Portugal, 25 September 1998.

6 Codner, "Strategic Defence Review," 7.

7 Barrie and Hichens, "Stealth Talks," 3.

8 Barrie, "British Air Chief Sceptical," 6.

9 *Strategic Defence Review*, par. 85.

10 Barrie, "UK Naval Emphasis Shifts," 16.

11 *Strategic Defence Review*, par. 110.

12 Ibid., par. 111.

13 Codner, "Strategic Defence Review," 8.

14 *Strategic Defence Review*, par. 109.

15 *Building Combat Capability*.

16 *Defence 2000*, 107.

17 Bostock, "Australian Maritime Missions," 5.

18 Evans, "The Middle Way," 5.

19 *Defence 2000*, 84.

20 Ibid., 78.

21 Ibid., 79.

22 Bostock, "RAAF."

23 *Defence 2000*, 79.

24 "Interview with Vice-Admiral D.J. Shackleton," 59.

25 Pearce, "Royal Australian Navy," 67.

26 Ferguson, "Race Begins," 14.

27 *Defence 2000*, xiii.

28 *An Australian Army*.

29 Géré, "RMA," 135.

30 Yost, *France*.

31 Douin, "Adapting French Defence," 2.

32 Mercier, "Les armées," 80.

33 Pengelley, "French Army in Profile," 40.

34 Jones, "French Forces," 34.

35 Yost, *France*, 10–11.

36 Foxwell, "France Weighs Up the Global Price," 30.

37 Mackenzie, "French Systems," 18.

38 Former French defence minister Pierre Joxe, as quoted ibid., 15.

39 As quoted in Thomas, *Military Challenges*, 53.

40 French general Allain Repplinger, deputy chief of staff for telecommunications and information systems, as quoted in Mackenzie, "France Sets Up Experimental Digitized Force," 12.

41 Jones, "French Forces," 35.

42 Boucheron, *Rapport*, as quoted in Grant, *The Revolution in Military Affairs*, 26.

43 "European UAVs Take Off," 39.

44 Mackenzie, "France Sets Up Experimental Digitized Force," 18.
45 Barrie, "Britain Splits," 1.
46 *Common Security.*
47 *White Paper 1994*, 16.
48 Thomas, *Military Challenges*, 24.
49 "Germany's New Look."
50 General Peter Carstens, chairman of the Commission on the Future of the Bundeswehr, as quoted in "A New Bundeswehr," 54.
51 Aguera, "German Navy," 12.
52 "European UAVs," 38.
53 Ibid.
54 "Shipborne UAVs," 40–1.

CHAPTER FIVE

1 "Knights in Shining Armour?" 3.
2 Fitchett, "U.S. Seeks More Defense Technology Cooperation," 6.
3 "Armies and Arms," 12.
4 Fitchett, "U.S. Seeks More Defense Technology Cooperation," 6.
5 The NATO Transatlantic Advanced Radar, which will provide the alliance with an airborne ground surveillance capability, was officially designated an alliance project in 2000. It is in the definition stage.
6 As quoted in Rogers, "Driving the Alliance?" 3.
7 Yost, "NATO Capabilities Gap," 106.
8 "Robertson's War," *The Times*, 5 August 1999.
9 Larrabee, *NATO's Adaptation*, 3.
10 Yost, "NATO Capabilities Gap," 119.
11 Tigner, "Focus May Shift," 22.
12 Andréani, Bertram, and Grant, *Europe's Military Revolution*, 55.
13 Ibid., 53–4.
14 Ibid., 56.
15 Robertson, "Can Europe Keep Up?" 49.
16 Former United Kingdom defence minister George Robertson, as quoted in Buerkle, "NATO Picks Briton," 1.
17 Andréani, Bertram, and Grant, 55.
18 Rogers, "Driving the Alliance," 3.
19 Gompert et al., *Mind the Gap*, 9.
20 "America versus Europe," 5.
21 Andréani, Bertram, and Grant, 53–4.
22 "NATO's Re-fit."
23 Robertson, "Can Europe Keep Up?" 49.
24 Former U.S. ambassador to NATO Robert Hunter, as quoted in Starr, "USA Warns," 4.

25 Clark, "Campaign in Kosovo," 6.

26 U.S. deputy defense secretary John Hamre, as quoted in Fitchett, U.S. Seeks More Cooperation, 6.

27 Rogers, "Driving the Alliance," 3.

28 Landay, "U.S. Military," 1.

29 Gompert et al., *Mind the Gap*, 4.

30 Schake et al., "Building a European Defence Capability," 21.

31 *Washington Summit Communiqué* (Washington, D.C.: NATO Heads of State and Government, April 1999), par. 24.

32 Thomas, *Military Challenges*, 66.

33 Tigner, "Focus May Shift," 10, 22.

34 Hill, "NATO Initiative," 2.

35 Ibid.

36 As quoted in "USA Sets More Realistic Goals," 74.

37 Yost, "NATO Capabilities Gap," 119.

38 Author interview with members of the U.S. mission to NATO, February 2000.

39 Thomas, "Military Challenges," 81.

CHAPTER SIX

1 Mazarr et al., *The Military Technical Revolution*, 49.

2 Vickers and Martinage, *The Military Revolution*, 27.

3 Daniel, *Coercive Inducement*, 70.

4 Morrocco, "Bombing Compels Serb Withdrawal," 36.

5 "Classified Report Highlights Benefits of PGMs," 1.

6 "Weather Was Toughest Problem."

7 Owen, "Aerospace Power," 20.

8 Department of Defense, *Kosovo*, 87.

9 Cordesman, *The Lessons*.

10 Department of Defense, *Kosovo*, 85.

11 *Aviation Week & Space Technology*, 20 September 1999, 25.

12 Fitchett, "NATO Lowers its Tally," 5.

13 Fulghum, "Pentagon Dissecting Kosovo Combat Data," 68.

14 Hillen, "Peacekeeping at the Speed of Sound," 7.

15 O'Hanlon, *Technological Change*, 129.

16 Miller, "U.K. Eyes Precision Weapons," 1–2.

17 Nye and Owens, "America's Information Edge," 24.

18 Gliksman, *Meeting the Challenge*, 9.

19 Howe, "Peacekeeping," 44.

20 Mazarr et al., *The Military Technical Revolution*, 50.

21 Holzer, "Military Trends," 8.

22 Defense Science Board, *Tactics and Technology*.

23 Metz and Kievit, *Revolution in Military Affairs.*

24 Garvelink, "Complex Emergencies," 67.

25 Fainberg, "Overview," 108.

26 Fiorenza, "Balkans Lessons," 74.

27 Anselmo, "Satellite Data," 29.

28 Thomas, *Virtual Peacemaking.*

29 "Prototype System," 1.

30 Becker, "They're Unmanned."

31 Whitney, "NATO Reconnaissance."

32 "NATO's Weapons," 47.

33 Thomas, "Kosovo."

34 Fulghum and Wall, "Joint-STARS," 74.

35 Proctor, "Aging NATO AWACS," 43.

36 Mazarr et al., *The Military Technical Revolution*, 51.

37 O'Hanlon, *Technological Change*, 131.

38 Graham and Priest, "Professional Consensus," 8.

39 Tirpak, "With Stealth in the Balkans," 24.

40 *Strategic Assessment 1999.*

41 Owen, "Aerospace Power," 12.

42 O'Hanlon, *Technological Change*, 131–3.

43 Pirnie, *Assessing Requirements for Peacekeeping,* 2.

44 Sengupta, "Heavyweight Tanks."

45 Fiorenza, "Balkans Lessons," 74.

46 See Shelton, "Peace Operations."

47 O'Hanlon, *Technological Change*, 130.

48 Scott, "Learning the Maritime Lessons"; Bostock, "RAAF Eyes A400M."

49 Willingham, "U.S. Air Force."

50 "Apache Flexes Reconnaissance Capabilities," 1.

51 Owen, "Aerospace Power," 10.

52 Mazarr et al., *The Military Technical Revolution*, 54.

53 Sen, "Wider Missions Emerge."

CHAPTER SEVEN

1 *1999 Joint Strategy Review.*

2 McKenzie Jr, *The Revenge of the Melians,* 2.

3 Schwartau, "Asymmetrical Adversaries," 203.

4 Sopko, "The Changing Proliferation Threat," 5–6.

5 Garrett, "The Nightmare of Bioterrorism," 76.

6 Sokolski, "Rethinking Bio-Chemical Dangers," 120.

7 Loeb and Anderson, "Al Qaeda."

8 Hogendoorn, "A Chemical Weapons Atlas," 36.

9 Sopko, "The Changing Proliferation Threat," 8.

10 Buchan, "One-and-a-Half Cheers."

11 Pfaltzgraff Jr, "Emerging Operational Challenges."

12 Bender, "Any Country Can Have an ICBM."

13 *Information Operations*, 2–2.

14 CIA director George Tenet, testimony before Congress, as stated in Pasternak, "Terrorism at the Touch of a Keyboard," 37. See also Bronskill, "Crippling Cyberattack."

15 Soo Hoo et al., "Information Technology," 138.

16 McKenzie Jr, *The Revenge of the Melians*, 60.

17 WMD proliferation expert with global security.org, Washington, DC, as quoted in "Intelligence Knew."

18 Mr Tony Kellet, defence analyst, Directorate of Strategic Analysis, Department of National Defence, 12 September 2001.

19 Starr, "War Games," 17.

20 Miller and Broad, "Clinton Describes Terrorism Threat."

21 Bloom and Eggen, "Airborne Poison."

22 CIA director George Tenet, as quoted in "Cyberwar Threat," 78.

23 Defence Science Advisory Board, *Information Operations*.

24 Garamone, "Joint Staff Releases Information Operations Doctrine."

25 Tiboni, "Attacks on DOD Computers," 10.

26 Seffers, "Report," 17.

27 Hum, "Defending Canadian Cyberspace."

28 *Critical Infrastructure Protection*, 2.

29 *Information Operations*, 1–8.

30 "Information Warfare and Cyberspace Security."

31 Munro, "Pentagon's New Nightmare."

32 Such as U.S. attorney general John Ashcroft. See "U.S. Fears Second Wave."

33 Seffers, "Report," 17.

34 Ibid.

35 "Bush Adds Clinton Holdover."

36 Canahute, "U.S. Military."

37 Donley, *Developing the Unified Command Plan*.

38 Canahute, "U.S. Military."

39 Sallot, "Guarding Canada's e-Frontier."

40 Fife, "Manley Will Run Security Sweep."

41 Bryden, "Chretien Announces Terrorism Crackdown."

42 Hobson, "The Asymmetric Future," 27.

43 Fife, "Push for Perimeter Begins."

CHAPTER EIGHT

1 Art Eggleton, keynote address to the conference Creating the Canadian forces of 2020, Ottawa, 26 April 2000.

2 General Maurice Baril, speech to the Canadian forces Command and Staff College, Toronto, 26 May 1999.

3 Air Force Facts & Stats.

4 Ibid.

5 Pugliese, "The U.S. Takes the Arms Race Sky-High."

6 Williams, "The Canadian Navy," 116.

7 Hobson and Lok, "Canadian Navy."

8 Message from Rear-Admiral D. Morse, commander, Standing Naval Force Atlantic, 28 March 2000.

9 "Army Regrouping in Face of Cuts," 29.

10 Stocker, "Canadian 'Jointery,'" 117.

11 Boomer, "Joint or Combined Doctrine?"

12 Stocker, "Canadian 'Jointery,'" 117.

13 Jockel, *The Canadian forces*, 73.

14 Dewar, "Revolution in Military Affairs."

15 Pugliese, "Canadians Will Link Service Info Systems."

16 *Chief of Air Staff Planning Guidance 2000*, 28.

17 Pugliese, "Analysts Skeptical," 12.

18 "Army Regrouping," 29.

19 Knight, "Army Takes Delivery."

20 Pugliese, "$140-Million Upgrade."

21 Jockel, *The Canadian Forces*, 54–5.

22 Brewster, "Ottawa Considers Investment."

23 Pugliese, "Canada Needs Stronger North American Defenses," 4.

24 See *Canada's Foreign Policy*.

25 Ibid., 13.

26 Interview with members of the U.S. mission to NATO, February 2000, as well as conversations with members of the U.S. defence community, on the margins of high-level military conferences, Washington, DC, 1999, and Boston, 1998.

27 Dewar, "Revolution in Military Affairs."

28 *Equipping and Modernizing the Canadian forces*.

29 Jockel, *The Canadian forces*.

30 Blanchard, "Canada Can't Keep Pace."

31 *1994 Defence White Paper*, 13.

32 See, for example, Trickey, "Canada's Fading Influence"; Buchanan, "World's New Power Group."

33 Maloney and Roberston, "The Revolution in Military Affairs." 460.

34 Gongora, "Canadian Land Forces 21."

EPILOGUE

1 Fukuyama, "The End of History?"

2 Huntington, "The Clash of Civilizations?"

3 Kaplan, "The Coming Anarchy."
4 Warren, "U.S. Forces"; Cordesman, *Lessons.*
5 "Send in the Drones," 73.
6 Eliot Cohen, as quoted in Ricks, "U.S. Arms Unmanned Aircraft."

Glossary

ADF	Australian Defence Force
APC	armoured personnel carrier
ARRC	Allied Command Europe Rapid Reaction Corps
AWACS	Airborne Warning and Control System
C4I	command, control, communications, computers, and intelligence
CALCM	conventional air-launched cruise missile
CJTF	Combined Joint Task Force
DCI	Defence Capabilities Initiative
DND	Department of National Defence
EPC	Emergency Preparedness Canada
FBI	Federal Bureau of Investigations
FEMA	Federal Emergency Management Agency
GPS	Global Positioning System
ISR	Intelligence, Surveillance, and Reconnaissance
JDAM	Joint Direct Attack Munition
JROC	Joint Requirements Oversight Council
JSTARS	Joint Surveillance Target Attack Radar System
MARCOT	Maritime Coordinated Operational Training
NMD	National Missile Defense
PGM	Precision Guided Munition
QDR	Quadrennial Defense Review
RCMP	Royal Canadian Mounted Police
RMA	revolution in military affairs
SDR	Strategic Defence Review
UAV	Unmanned Aerial Vehicle

| UCAV | Unmanned Combat Aerial Vehicle |
| WMD | weapons of mass destruction |

Bibliography

1994 Defence White Paper. Ottawa: Department of National Defence, December 1994.

1999 Joint Strategy Review. Washington, DC: Joint Chiefs of Staff.

2001–2002 Report on Plans and Priorities. Ottawa: Department of National Defence.

Adams, Gordon. "British Set Example of Focused Strategic Defense Review." *Defence News*, 17–23 August 1998.

Adjusting Course: A Naval Strategy for Canada. Ottawa: Department of National Defence, April 1997.

Aguera, Martin. "German Navy Working to Be a More Flexible, Mission-Ready Force." *Defense News*, 30 October 2000.

Air Force Facts & Stats, http: //www.airforce.dnd.ca.

Allen, Mike. "Bush Vows Spending on Futuristic Weapons." *Washington Post*, 13 February 2001.

"America versus Europe." *Economist*, 24 April 1999.

America's Air Force: Global Vigilance, Reach and Power. Washington, DC: Chief of Staff of the Air Force, 2000.

Andréani, Gilles, et al. *Europe's Military Revolution*. London: Centre for European Reform, March 2001.

Anselmo, Joseph. "Satellite Data Plays Key Role in Bosnia Peace Treaty." *Aviation Week & Space Technology*, 11 December 1995.

"Apache Flexes Reconnaissance Capabilities in Bosnia." *Tactical Technology*, 21 February 1996.

"Armies and Arms." *Economist*, 24 April 1999.

"Army Regrouping in Face of Cuts." *Jane's Defence Weekly*, 2 February 2000.

Army Vision 2010. Washington, DC: Chief of Staff of the Army, November 1996.

Auditor General of Canada. *Equipping and Modernizing the Canadian Forces*. Ottawa: April 1998

Australian Army for the Twenty-first Century. An. Canberra: Department of Defence, October 1996.

Australia's Strategic Policy. Canberra: Department of Defence, December 1997.

Baril, Maurice. Chief of Defence Staff, Canada. Speech to the Canadian Forces Command and Staff College, Toronto, 26 May 1999.

Barrie, Douglas. "Britain May Speed Up UAV Project." *Defense News*, 28 May to 3 June 2001.

–"Britain Splits from Europe on Stealth Effort." *Defense News*, 9 October 2000.

–"British Air Chief Sceptical about Feasibility of UCAV." *Defense News*, 30 November to 6 December 1998.

–"UK Naval Emphasis Shifts to Shallow Waters." *Defense News*, 16–22 November 1998.

Barrie, Douglas, and Theresa Hichens. "Stealth Talks May Hold Key to Britain's FOAS." *Defense News*, 31 August 1998.

Becker, Elizabeth. "They're Unmanned, They Fly Low, and They Get the Picture." *New York Times*, 3 June 1999.

Bender, Bryan. "Any Country Can Have an ICBM by 2015, says CIA." *Jane's Defence Weekly*, 5 May 1999.

–"Budget Pressures Are Trapping U.S. Forces 'In A Death Spiral.'" *Jane's Defence Weekly*, 16 September 1998.

–"DOD Expands Pursuit of Interoperability." *Jane's Defence Weekly*, 29 September 1999.

Biddle, Stephen. "Victory Misunderstood: What the Gulf War Tells Us about the Future of Conflict." *International Security* (fall 1996).

Blaker, James R. *Understanding the Revolution in Military Affairs.* Washington, DC: Progressive Policy Institute, January 1997.

Blanchard, Mike. "Canada Can't Keep Pace with U.S. Technology: Baril." *Ottawa Citizen*, 4 May 2001.

Blank, Stephen J. "The Illusion of a Short-War." *SAIS Review* (winter-spring 2000).

–"Preparing for the Next War: Reflections on the Revolution in Military Affairs." *Strategic Review* (spring 1996).

Bloom, Justin, and Dan Eggen. "Airborne Poison via Crop Duster?" *International Herald Tribune*, 25 September 2001.

Boomer, F.M. "Joint or Combined Doctrine? The Right Choice for Canada." Paper submitted to the Advanced Military Studies Course, Canadian Forces Command and Staff College, Toronto, 1998.

Bostock, Ian. "Australian Maritime Missions Await Global Hawk UAV." *Jane's Defence Weekly*, 2 May 2001.

–"RAAF Eyes A400M for Heavy Airlift Role." *Jane's Defence Weekly*, 7 March 2001.

Boucheron, Jean-Michel. *Rapport Fait au Nom de la Commission des Finances.* Paris: Assemblé Nationale, November 1997.

Brewster, Murray. "Ottawa Considers Investment to Develop Stealth Fighter Jet." *Globe and Mail*, 12 March 2001.

Bronskill, Jim. "Crippling Cyberattack in 10 Years: CSIS." *Ottawa Citizen*, 18 July 2001.

Bryden, Joan. "Chrétien Announces Terrorism Crackdown." *Ottawa Citizen*, 2 October 2001.

Buchan, Glenn C. "One-and-a-Half Cheers for the Revolution in Military Affairs." Paper presented to the conference Defence and Security at the Dawn of the Twenty-first Century, Sainte-Foy, Quebec, 2–4 October 1997.

Buchanan, Carrie. "World's New Power Group Doesn't Include Canada." *Ottawa Citizen*, 25 November 1999.

Buerkle, Tom. "NATO Picks Briton as Its Next Leader." *International Herald Tribune*, 5 August 1999.

Building Combat Capability. Canberra: Department of Defence, 1998.

"Bush Adds Clinton Holdover to Security Team." *Washington Times*, 1 October 2001.

Bush, George W. "A Period of Consequences." Speech to The Citadel, 23 September 1999.

Canada's Army (Ottawa: Department of National Defence, Canadian Forces Publication B-GL-300–000/FP-000, April 1998).

Canada's Foreign Policy: Principles and Priorities for the Future. Ottawa: Report of the Special Joint Committee of the Senate and the House of Commons, reviewing Canadian foreign policy, November 1994.

Canahuate, Tom. "Pentagon to Accelerate Development of UAVs, Precision Munitions." *Defense News*, 24 September 2001.

–"U.S. Army Considers Acceleration of FCS Program." *Defense News*, 26 September 2001.

–"U.S. Military Welcomes New Homeland Security Position." *Defense News*, 25 September 2001.

Chief of Air Staff Planning Guidance 2000. Ottawa: Department of National Defence, November 1999.

Clark, Colin. "Campaign in Kosovo Highlights Allied Interoperability Shortfalls." *Defense News*, 16 August 1999.

–"U.S. Army Alters Schedule for Digitization System." *Defense News*, 16 August 1999.

"Classified Report Highlights Benefits of PGMs: NATO Air Forces Inflicted No Collateral Damage in Bosnian Air Campaign." *Inside the Air Force*, 19 July 1996.

Codner, Michael. "The Strategic Defence Review: How Much? How Far? How Joint Is Enough?" *RUSI Journal* (August 1998).

Cohen, Eliot A. "Defending America in the Twenty-first Century." *Foreign Affairs* (November-December 2000).

–"A Revolution in Warfare." *Foreign Affairs* (March-April 1996).

Cohen, William S. *Annual Report to the President and the Congress.* Washington, DC: 2001.

Common Security and the Future of the Bundeswehr. Berlin: Report of the Commission to the Federal Government, May 2000.

Cook, Nick. "UK to Accelerate Watchkeeper UAV Program." *Jane's Defence Weekly,* 1 November 2000.

Cooper, Pat. "U.S. Stealth Enhancements are Key to 'Air Occupation.'" *Defense News,* 16–22 September 1996.

Cordesman, Anthony. *The Lessons and Non-Lessons of the Air and Missile War in Kosovo.* Washington, DC: Center for Strategic and International Studies, July 1999.

Correll, John T. "Aerospace Power Meets the QDR." *Air Force Magazine* (July 2000).

"Cyberwar Threat Real and Growing." *Aviation Week & Space Technology,* 27 April 1998.

Daniel, Donald C.F. *Coercive Inducement and the Containment of International Crises.* Washington, DC: United States Institute of Peace Press, 1998.

Davis, Jacquelyn K., et al. *The Submarine and U.S. National Security Strategy into the Twenty-first Century.* Cambridge, MA: Institute for Foreign Policy Analysis, National Security Studies Paper no. 19, 1997.

Davis, Norman. "An Information-Based Revolution in Military Affairs." *Strategic Review* (winter 1996).

Defence 2000: Our Future Defence Force. Canberra: Department of Defence, December 2000. *Defence Plan 2001.* Ottawa: Department of National Defence, April 2001.

Defence Planning Guidance 2000. Ottawa: Department of National Defence, August 1999.

Defence Planning Guidance 2001. Ottawa: Department of National Defence, April 2000.

Defence Program Law. Paris: Ministry of Defence, July 1996.

Defence Science Advisory Board. *Information Operations.* Ottawa: Department of National Defence, 1997.

Defense Science Board. *Tactics and Technology for Twenty-first Century Military Superiority.* Washington, DC: Office of the Secretary of Defense, 1996.

Department of Defense. *Kosovo/Operation Allied Force after Action Report.* Washington, DC: Report to Congress, January 2000.

Dewar, J.S. "Revolution in Military Affairs: The Divergence between the Most Dangerous and the Most Likely." Paper submitted to the Advanced Military Studies Course, Canadian Forces Command and Staff College, Toronto, 1998.

Donley, Michael B. *Developing the Unified Command Plan for "Homeland Security".* Washington, DC: Report to the Joint Staff Directorate for Strategic Plans and Policy, February 2001.

Donnelly, Tom. "Revolution? What Revolution?" *Jane's Defence Weekly*, 7 June 2000.

Douin, Jean-Philippe. "Adapting French Defence to the New Geostrategic Context." *RUSI Journal* (August 1997).

Eggleton, Art. Defence Minister, Canada. Keynote Address to the Conference Creating the Canadian Forces of 2020, Ottawa, 26 April 2000.

"European UAVs Take Off." *Armed Forces Journal International* (July 1999).

Evans, Michael. "The Middle Way." *National Security Studies Quarterly* (winter 2000).

Evers, Stacey. "USN's Next Generation Destroyer to be Stealthy and Smart." *Defense News*, 22 October 1997.

Fainberg, Anthony. "Overview of Key Technologies for Peace Operations." In Alex Gliksman, ed., *Meeting the Challenge of International Peace Operations: Assessing the Contribution of Technology*. Livermore, CA: Lawrence Livermore National Laboratory Center for Global Security Research, 1998.

Ferguson, Gregor. "Race Begins for Australian Fighter Aircraft Replacement." *Defense News*, 29 November 1999.

Fife, Robert. "Manley Will Run Security Sweep." *National Post*, 2 October 2001.

–"Push for Perimeter Begins." *National Post*, 19 September 2001.

Fiorenza, Nicholas. "Balkans Lessons: European Ground Forces Become Leaner, but Are They Meaner?" *Armed Forces Journal International* (June 2000).

Fitchett, Joseph. "NATO Lowers its Tally of Tank Hits in Kosovo," *International Herald Tribune*, 17 September 1999.

–"US Seeks More Defense Technology Cooperation with Europeans," *International Herald Tribune*, 14 June 1999.

Flournoy, Michèle A. *Report of the National Defense University Quadrennial Defense Review 2001 Working Group*. Washington, DC: Institute for National Strategic Studies, November 2000.

Forward ... From the Sea. Washington, DC: Chief of Naval Operations, 1994.

Foxwell, David. "France Weighs up the Global Price." *Jane's Navy International*, 1 July 1998.

Foxwell, David, and Joris Janssen Lok. "Approaching Vanishing Point: The Emergence of Stealth Ships." *Jane's International Defense Review* (9/1998).

Friedman, George, and Meredith Friedman. *The Future of War*. New York, NY: Crown Publishers, 1996.

Friesen, Shaye K., ed. *Transforming an Army: Land Warfare Capabilities for the Future Army*. Kingston, ON: Directorate of Land Strategic Concepts, July 1999.

Fukuyama, Francis. "The End of History?" *National Interest* (summer 1989).

Fulghum, David A. "Pentagon Dissecting Kosovo Combat Data." *Aviation Week & Space Technology*, 26 July 1999.

Fulghum, David A., and Robert Wall. "Joint-STARS May Profit from Yugoslav Ops." *Aviation Week & Space Technology*, 26 July 1999.

"Future of Warfare, The." *Economist*, 8 March 1997.

Garamone, Jim. "Joint Staff Releases Information Operations Doctrine." *American Forces Press Service*, 10 March 1999.

Garrett, Laurie. "The Nightmare of Bioterrorism." *Foreign Affairs* (January-February 2001).

Garvelink, William J. "Complex Emergencies in Africa in the 1990s: The Role of Technology." In Alex Gliksman, ed., *Meeting the Challenge of International Peace Operations: Assessing the Contribution of Technology*. Livermore, CA: Lawrence Livermore National Laboratory Center for Global Security Research, 1998.

General Accounting Office. *Critical Infrastructure Protection*. Washington, DC: June 2000.

Géré, Francois. "RMA or New Operational Art? A View From France." In Thierry Gongora and Harald von Riekhoff, eds., *Toward a Revolution in Military Affairs? Defense and Security at the Dawn of the Twenty-first Century*. Westport, CT: Greenwood Press, 2000.

"Germany's New Look Security Policy." *IISS Strategic Comments* (January 1997).

Gertz, Bill. "Rumsfeld Recalls Cold War." *Washington Times*, 5 October 2001.

Gliksman, Alex, ed. *Meeting the Challenge of International Peace Operations: Assessing the Contribution of Technology*. Livermore, CA: Lawrence Livermore National Laboratory Center for Global Security Research, 1998.

Global Engagement: A Vision for the Twenty-first Century Air Force. Washington, DC: Chief of Staff of the Air Force, 1996.

Gompert, David C., et al. *Mind the Gap: Promoting a Transatlantic Revolution in Military Affairs*. Washington, DC: National Defense University Institute for National Strategic Studies, March 1999.

Gongora, Thierry. "Canadian Land Forces 21: A Multi-Purpose Force for the Next Century." Paper presented to the Conference of Defence Associations Institute Canadian Military Affairs Symposium, Ottawa, 13–14 November 1998.

Gongora, Thierry, and Harald von Riekhoff, eds. *Toward a Revolution in Military Affairs? Defense and Security at the Dawn of the Twenty-first Century*. Westport, CT: Greenwood Press, 2000.

Goodman, Glenn W. "Aerospace Force." *Armed Forces Journal International* (September 2000). Gouré, Daniel. "Is There a Military-Technical Revolution in America's Future?" *Washington Quarterly* (autumn 1993).

–"The Resource Gap." *Armed Forces Journal International* (May 2000).

Graham, Bradley, and Dana Priest. "Professional Consensus: No Way to Fight a War." *International Herald Tribune*, 7 June 1999.

Grant, Robert P. *The Revolution in Military Affairs and European Defense Cooperation.* St Augustin, Germany: Konrad Adenauer Foundation, Arbeitspapier, June 1998.

Henley, Lonnie D. "The RMA after Next." *Parameters* (winter 1999–2000).

Hewish, Mark. "Building a Bird's-Eye View of the Battlefield." *Jane's International Defense Review* (2/1997).

– "Coming Soon: Attack of the Killer UAVs," *Jane's International Defense Review* (9/1999).

Hill, Luke. "NATO Initiative Progress Lags." *Jane's Defence Weekly,* 30 May 2001.

Hillen, John. "Peacekeeping at the Speed of Sound: The Relevancy of Airpower Doctrine in Operations Other than War." *Airpower Journal* (winter 1998).

Hobson, Sharon. "The Asymmetric Future." *Jane's Defence Weekly,* 23 August 2000.

Hobson, Sharon, and Joris Janssen Lok. "Canadian Navy Moves to Fill C4I Gap." *Jane's International Defense Review,* 1 June 1999.

Hogendoorn, E.J. "A Chemical Weapons Atlas." *Bulletin of the Atomic Scientists* (September-October 1997).

Holzer, Robert. "Experts Warn U.S. Lacks Strategy for Change." *Defense News,* 28 August 2000.

– "Military Trends Demand More Complex Weapons." *Defense News,* 25 October 1999.

– "Questions Swirl around Air Campaign." *Defense News,* 12 April 1999.

– "Think Tank Defense Vision Targets Big Ticket Projects." *Defense News,* 7 August 2000.

– "Transformation Must Leap Many Hurdles." *Defense News,* 19 March 2001.

– "U.S. Navy Sees Initial Impact of Network-Centric Warfare." *Defense News,* 12–18 October 1998.

– "U.S. Navy Steps Up Data-Linked Warfare Pace." *Defense News,* 10–16 November 1997.

Howe, Jonathan T. "Peacekeeping and Peace Enforcement: Can Technology Help? Lessons from UNOSOM II." In Alex Gliksman, ed. *Meeting the Challenge of International Peace Operations: Assessing the Contribution of Technology.* Livermore, CA: Lawrence Livermore National Laboratory Center for Global Security Research, 1998.

Hum, Peter. "Defending Canadian Cyberspace: DND Seeks to Co-ordinate Battle Plan with Spy Agency and Mounties." *Ottawa Citizen,* 22 March 1997.

Hundley, Richard O. *Past Revolutions, Future Transformations: What Can the History of Revolutions in Military Affairs Tell Us about Transforming the U.S. Military?* Santa Monica, CA: RAND, 1999.

Huntington, Samuel. "The Clash of Civilizations?" *Foreign Affairs* (summer 1993).

Information Operations. Washington, DC: Chief of Staff of the Army, United States Army Field Manual 100–6, August 1996.

"Information Warfare and Cyberspace Security." *RAND Research Review* (fall 1995).

"Intelligence Knew 'Some Group Planning This.'" *National Post Online,* 12 September 2001.

"Interview with Vice-Admiral D.J. Shackleton." *U.S. Naval Institute Proceedings* (January 2000).

Jockel, Joseph T. *The Canadian Forces: Hard Choices, Soft Power.* Toronto: Canadian Institute of International Affairs, 1999.

Joint Vision 2010. Washington, DC: Joint Chiefs of Staff, June 1996.

Joint Vision 2020. Washington, DC: Joint Chiefs of Staff, May 2000.

Jones, Jeffrey. "French Forces for the Twenty-first Century." *Joint Forces Quarterly* (summer 2000).

Kaplan, Robert. "The Coming Anarchy," *Atlantic Monthly* (February 1994).

"Keep Pace with Technology." *Defense News,* 3–9 August 1998.

Knight, Sue. "Army Takes Delivery of First LAV III's." *Maple Leaf,* 8 September 1999.

"Knights in Shining Armour?" *Economist,* 24 April 1999.

Krepinevich, Andrew. "Cavalry to Computer: The Pattern of Military Revolutions." The *National Interest* (fall 1994).

–*Wither the Army.* Washington, DC: Center for Strategic and Budgetary Assessments, January 2000.

Lambeth, Benjamin S. "The Technology Revolution in Air Warfare." *Survival* (spring 1997).

Landay, Jonathan S. "U.S. Military Outpaces its NATO Peers." *Christian Science Monitor,* 29 September 1997.

Larrabee, F. Stephen. *NATO's Adaptation and Transformation: Key Challenges.* Santa Monica, CA: RAND. Statement before the Senate Foreign Relations Committee, 21 April 1999.

Lewis, Leslie, and C. Robert Roll. "Quadrennial Defense Review 2001: Managing Change in the Department of Defence." *National Security Studies Quarterly* (autumn 2000).

Loeb, Vernon, and John Ward Anderson. "Al Qaeda May Have Crude Chemical, Germ Capabilities." *Washington Post,* 27 September 2001.

Luttwak, Edward N. "A Post-Heroic Military Policy." *Foreign Affairs* (July-August 1996).

Macgregor, Douglas. *Breaking the Phalanx: A New Design for Landpower in the Twenty-first Century.* Westport, CT: Praeger Publishers, 1997.

Mackenzie, Christina. "France Sets Up Experimental Digitized Force." *Defense News,* 29 May 2000.

–"French Systems to Reflect Mission Changes." *Defense News,* 30 October 2000.

Maloney, Sean, and Scot Roberston. "The Revolution in Military Affairs: Possible Implications for Canada." *International Journal* (summer 1999).

Marine Corps Strategy 21. Washington, DC: Commandant of the Marine Corps, November 2000.

Marshall, Andrew. Director of the Office of Net Assessment, Office of the Secretary of Defense. Statement before the Senate Armed Services Committee, 5 May 1995.

Mastanduno, Michael. "Preserving the Unipolar Moment." *International Security* (spring 1997).

Matthews, William. "Improved Space Assets Would Aid U.S. Forces in Persian Gulf Conflict." *Defense News*, 23 February to 1 March 1998.

Mazarr, Michael J., et al. *The Military Technical Revolution.* Washington, DC: Center for Strategic and International Studies, March 1993.

McGinn, John G. "Lessons from the National Defense Panel." *Joint Forces Quarterly* (spring 2000).

McKenzie, Kenneth F. Jr. *The Revenge of the Melians: Asymmetric Threats and the Next QDR.* Washington, DC: National Defense University Institute for National Strategic Studies, McNair Paper no. 62, November 2000.

Mercier, Philippe. "Les armées et leurs équipements." *Défense nationale* (July 1996).

Metz, Steven, and James Kievit. *The Revolution in Military Affairs and Conflict Short of War.* Carlisle, PA: United States Army War College Strategic Studies Institute, July 1994.

Miller, Charles. "U.K. Eyes Precision Weapons for Roles in Peacekeeping." *Defense News*, 24–30 June 1996.

Miller, Judith, and William J. Broad. "Clinton Describes Terrorism Threat for Twenty-first Century." *New York Times*, 22 January 1999.

Morrocco, John D. "Bombing Compels Serb Withdrawal." *Aviation Week & Space Technology*, 25 September 1995.

Morse, D., Commander Standing Naval Force Atlantic. Message dated 28 March 2000.

Munro, Neil. "The Pentagon's New Nightmare: An Electronic Pearl Harbor." *Washington Post*, 16 July 1995.

Murray, Williamson. "Thinking about Revolutions in Military Affairs." *Joint Forces Quarterly* (summer 1997).

National Defense Panel. *Transforming Defense: National Security in the Twenty-first Century* (Washington, DC: December 1997).

NATO Heads of State and Government. *Washington Summit Communiqué*, Washington, DC, 24 April 1999.

"NATO's Re-fit: Big Changes Before the April Summit?" *Jane's Foreign Report*, 29 October 1998.

"NATO's Weapons: Are They Too Clever By Half?" *Economist*, 1 May 1999.

Navy Operational Concept. Washington, DC: Chief of Naval Operations, March 1997).

"The New Air Force Line-Up 'Lighter, Leaner and More Lethal.'" *Jane's Defence Weekly*, 8 September 1999.

"New Arms for a New World." *Economist*, 17 February 2001.

"New Bundeswehr." *Armed Forces Journal International (June 1999)*.

Nye, Joseph S., and William A. Owens. "America's Information Edge." *Foreign Affairs* (March-April 1996).

O'Hanlon, Michael. "Can High Technology Bring U.S. Troops Home?" *Foreign Policy* (winter 1998–99).

–"Prudent or Paranoid? The Pentagon's Two-War Plans," *Survival* (spring 2001).

–*Technology Change and the Future of Warfare*. Washington, DC: Brookings Institution Press, 2000).

Operational Manoeuvre from the Sea. Washington, DC: Commandant of the Marine Corps.

Owens, Mackubin Thomas. "Technology, the RMA, and Future War." *Strategic Review* (spring 1998).

Owen, Robert C. "Aerospace Power and Land Power in Peace Operations." *Airpower Journal* (fall 1999).

Pasternak, Douglas. "Terrorism at the Touch of a Keyboard." *U.S. News & World Report*, 13 July 1998.

Pearce, Henry. "The Royal Australian Navy: On Course for the Twenty-first Century." *U.S. Naval Institute Proceedings* (March 2001).

Pengelley, Rupert. "French Army in Profile: From Hollow Force to Hard Core." *Jane's International Defence Review* (6/2000).

Perry, William. "Desert Storm and Deterrence." *Foreign Affairs* (fall 1991).

Pfaltzgraff, Robert Jr. "Emerging Operational Challenges in an Evolving Spectrum of Conflict." Paper presented to the conference on the Role of Naval Forces in Twenty-first Century Operations, Cambridge, Massachusetts, 19–20 November 1997.

Pirnie, Bruce. *Assessing Requirements for Peacekeeping, Humanitarian Assistance, and Disaster Relief*. Santa Monica, CA: RAND, 1998.

Proctor, Paul. "Aging NATO AWACS Prove Reliable in Balkans Campaign." *Aviation Week & Space Technology*, 30 August 1999.

"Prototype System Integrates Air Picture for Bosnia." *Tactical Technology*, 24 January 1996.

Pugliese, David. "$140-Million Upgrade Called a Waste," *National Post*, 23 November 1999.

–"Analysts Skeptical about Canada's Fast-Reaction Plans." *Defense News*, 14 May 2001.

–"Canada Needs Stronger North American Defenses." *Defense News*, 9 April 2001.

-"Canadians Will Link Service Info Systems." *Defense News*, 28 September 1998.

-"The u.s. Takes the Arms Race Sky-High." *Ottawa Citizen*, 13 January 2001.

Quadrennial Defense Review. Washington, DC: Department of Defense, May 1997.

Quadrennial Defense Review. Washington, DC: Department of Defense, September 2001.

"Remarks by the President to the Troops and Personnel." *White House Press Release*, 13 February 2001.

Ricks, Thomas E. "For Rumsfeld, Many Roadblocks." *Washington Post*, 7 August 2001.

-"u.s. Arms Unmanned Aircraft." *Washington Post*, 18 October 2001.

Robertson, George. "Can Europe Keep Up with the Revolution in Military Affairs?" *RUSI Journal* (April-May 1999).

Robertson, George. Defence Secretary, Britain. Intervention at NATO Defence Ministerial, Vilamoura, Portugal, 25 September 1998.

"Robertson's War." *The Times*, 5 August 1999.

"Robots Herald New Era." *Defense News*, 2 April 2001.

Rogers, Clifford. "'Military Revolutions' and 'Revolutions in Military Affairs': A Historian's Perspective." In Thierry Gongora and Harald von Riekhoff, eds. *Toward a Revolution in Military Affairs? Defense and Security at the Dawn of the Twenty-first Century.* Westport, CT: Greenwood Press, 2000.

Rogers, Marc. "Driving the Alliance? NATO Follows the u.s. Lead." *IDR Special Report*, 1 December 1998.

Roos, John G. "Disappearing Act: Ground Forces Embrace Stealth." *Armed Forces Journal International* (October 1998).

Sallot, Jeff. "Guarding Canada's e-Frontier." *Globe and Mail*, 20 February 2001.

Scarborough, Rowan. "Pentagon to Focus on Defense of u.s. Soil." *Washington Times*, 2 October 2001.

Schake, Kori, et al. "Building a European Defence Capability." *Survival* (spring 1999).

Schwartau, Winn. "Asymmetrical Adversaries." *Orbis* (spring 2000).

Science, Technology and Military Strength. Washington, DC: Center for Strategic and Budgetary Assessments, November 1999.

Scott, Richard. "Learning the Maritime Lessons of East Timor." *Jane's Defence Weekly*, 30 August 2000.

Seffers, George. "Land Forces Seek to Shorten Logistics Tail." *Defense News*, 20–26 April 1998.

-"Report: u.s. Needs to Avert Electronic Waterloo." *Defense News*, 11 January 1999.

–"U.S. Army Study: Reduce Force Logistics, Improve Mobility." *Defense News*, 16–22 December 1996.

Seigle, Greg. "USA Forms Joint Force Command to Defeat Threats." *Jane's Defence Weekly*, 6 October 1999.

Sen, Philip. "Wider Missions Emerge for Australia's Fleet Air Arm." *Jane's Navy International*, 1 March 2001.

"Send in the Drones." *Economist*, 10 November 2001.

Sengupta, Kim. "Heavyweight Tanks Are Defeated by Rivals." *Independent*, 14 April 2001.

Shaping the Future of the Canadian Forces: A Strategy for 2020. Ottawa: Department of National Defence, June 1999.

Shelton, Henry H. "Peace Operations: The Forces Required." *National Security Studies Quarterly* (summer 2000).

"Shipborne UAVs Await Take Off." *Jane's Navy International* (January–February 1998).

Snodgrass, David E. "The QDR: Improve the Process to Improve the Product." *Parameters* (spring 2000).

Sokolski, Henry. "Rethinking Bio-Chemical Dangers." *Orbis* (spring 2000).

Soldiers on Point for the Nation: Persuasive in Peace, Invincible in War. Washington, DC: Chief of Staff of the Army, October 1999.

Soo Hoo, Kevin, et al. "Information Technology and the Terrorist Threat." *Survival* (autumn 1997).

Sopko, John F. "The Changing Proliferation Threat." *Foreign Policy* (winter 1996–97).

Starr, Barbara. "USA Warns of Three-Tier NATO Technology Rift." *Defense News*, 1 October 1997.

–"War Games Highlight U.S. Vulnerability in Space." *Jane's Defence Weekly*, 8 October 1997.

Stocker, Jeremy. "Canadian 'Jointery.'" *Joint Forces Quarterly* (winter 1995–96).

Strategic Assessment 1999. Washington, DC: National Defense University Institute for National Strategic Studies.

Strategic Defence Review. London: Ministry of Defence, July 1998.

Thomas, James P. *The Military Challenges of Transatlantic Coalitions*. London: International Institute for Strategic Studies, Adelphi Paper no. 333, May 2000.

Thomas, Timothy L. "Kosovo and the Current Myth of Information Superiority." *Parameters* (spring 2000).

–*Virtual Peacemaking: A Military View of Conflict Prevention through the Use of Information Technology*. Fort Leavanworth, KS: Foreign Military Studies Office,1999.

Tiboni, Frank. "Attacks on DOD Computers Nearly Double 1999 Level." *Defense News*, 21 August 2000.

Tigner, Brooks. "Focus May Shift from Kosovo to EU Defense Force Plans." *Defense News*, 18 December 2000.

Tirpak, John A. "With Stealth in the Balkans." *Air Force Magazine* (October 1999).

Toffler, Alvin, and Heidi Toffler. *War and Anti-War*. New York, NY: Warner Books, 1993.

Trickey, Mike. "Canada's Fading Influence Linked to Military Cuts." *Ottawa Citizen*, 1 November 1999.

Tucker, Robert W. "The Future of a Contradiction." *National Interest* (spring 1996).

United States Commission on National Security/Twenty-first Century. *New World Coming: American Security in the Twenty-first Century*. Washington, DC: Phase I Report, September 1999.

–*Seeking a National Strategy: A Concert for Preserving Security and Promoting Freedom*. Washington, DC: Phase II Report, April 2000.

–*Road Map for National Security: Imperative for Change*. Washington, DC: Phase III Report, January 2001.

"The Unmanned Inevitability?" *Defense & Foreign Affairs Strategic Policy* (April-May 1998).

"U.S. Fears Second Wave of Terror." *Ottawa Citizen*, 1 October 2001.

"U.S. Military Spending: Policy at a Crossroads." *Strategic Comments*. London: International Institute for Strategic Studies, May 2000.

"USA Sets More Realistic Goals for NATO's DCI." *Jane's Defence Weekly*, 14 June 2000.

"USA Takes Latest C2 System into the Field." *Jane's Defence Weekly*, 18 February 1998.

Vick, Alan, et al. *Preparing the U.S. Air Force for Military Operations Other than War*. Santa Monica, CA: Rand, 1997.

Vickers, Michael G., and Robert C. Martinage. *The Military Revolution and Intrastate Conflict*. Washington, DC: Center for Strategic and Budgetary Assessments, October 1997.

Walsh, Mark. "USAF Targets Control in Air, Space." *Defense News*, 21–27 April 1997.

Warren, David. "U.S. Forces Not Just Bigger, They're Better." *Ottawa Citizen*, 12 December 2001.

Watkins, Steven. "USAF System Links Battlefield Intelligence Data." *Defense News*, 12–18 August 1997.

"Weather Was Toughest Problem in Bosnia." *Tactical Technology*, 4 October 1995.

White Paper 1994: The Security of the Federal Republic of Germany and the Bundeswehr Now and in the Years Ahead. Bonn: Ministry of Defence, April 1994.

Whitney, Craig R. "NATO Reconnaissance Fails to Locate Missing Refugees." *New York Times*, 10 April 1999.

Why No Transformation? Washington, DC: Center for Strategic and Budgetary Assessments, February 1999.

Williams, Kelly. "The Canadian Navy: Ready and Relevant." In Shaye K. Friesen, ed., *Transforming an Army: Land Warfare Capabilities for the Future Army*. Kingston, ON: Directorate of Land Strategic Concepts, July 1999.

Willingham, Stephen. "U.S. Air Force Touts Success of African Flood Relief Effort." *National Defense* (May 2000).

Willis, G.E. "Advances Will Expand Role, Power of Field Artillery." *Defense News*, 16–22 February 1998.

Wirtz, James J. "QDR 2001: The Navy and the Revolution in Military Affairs." *National Security Studies Quarterly* (autumn 1999).

"Working Out the World." *Economist*, 31 March 2001.

Yost, David S. *France and the Revolution in Military Affairs*. Unpublished manuscript, 31 March 1997.

–"The NATO Capabilities Gap and the European Union." *Survival* (winter 2000, 2001).

Index